Historia Logicae
Volume 1

Historia Logicae and its Modern Interpretation

Volume 1
Historia Logicae and its Modern Interpretation
Jens Lemanski and Ingolf Max, eds.

Historia Logicae series editors
Jens Lemanski					jens.lemanski@fernuni-hagen.de
Ingolf Max					max@uni-leipzig.de

Historia Logicae and its Modern Interpretation

Editors

Jens Lemanski

Ingolf Max

© Individual author and College Publications 2023
All rights reserved.

ISBN 978-1-84890-408-8

College Publications
Scientific Director: Dov Gabbay
Managing Director: Jane Spurr

http://www.collegepublications.co.uk

Original cover design by Laraine Welch

All rights reserved. No part of this publication may be reproduced, stored in a retrieval system or transmitted in any form, or by any means, electronic, mechanical, photocopying, recording or otherwise without prior permission, in writing, from the publisher.

Table of Contents

PREFACE – HISTORIA LOGICAE AND ITS MODERN INTERPRETATION   vii

I   ANCIENT AND MEDIEVAL LOGIC   1

1. Ioannis M. Vandoulakis – Self-Reference and type Distinctions in Greek Philosophy and Mathematics   3

2. Victor Nascimento – Stoic Natural Deduction and Intuitionistic Stoic Logic   37

3. Jens Lemanski – Seneca's and Porphyry's Trees in Modern Interpretation   61

4. Manuel Correia – The Number of Categorical Propositions in Boethius's Commentaries on Aristotle's Peri Hermeneias   89

5. Manuel Dahlquist and Luis A. Urtubey – Inferring Without Truth: Revisiting some Buridan's Puzzling Inferences   107

6. Joshua Mendelsohn – "Men go grey": Robert Kilwardby and the Logic of Natural Contingency   145

II   MODERN AGE LOGIC   189

7. Wolfgang Lenzen – Caramuel's Pentagon of Opposition and his Vindication of the Principle Ex Contradictorio Quodlibet   191

8 Valentin Pluder – Formal Logic Between Metaphysics and Purpose: Three Different Notions of ‚Formal Logic' in the Early 19th Century   213

9 Víctor Aranda – Non-Denoting Terms and "Imaginary" Numbers: Modern Interpretations of Husserl's 1901 Double Lecture   231

10 Ryan Christensen – The Model-Theoretic Square of Opposition   251

11 Fabio De Martin Polo – Discussive Logic: A Short History of the First Paraconsistent Logic   267

Historia Logicae and Its Modern Interpretation

Jens Lemanski
University of Münster, University of Hagen, Germany
`jens.lemanski@fernuni-hagen.de`

Ingolf Max
University of Leipzig, Germany
`max@rz.uni-leipzig.de`

With the innovations within the logic of the late 19th and early 20th century, the history of logic was long regarded as an outdated discipline that could no longer contribute to current developments and ideas. The gap between traditional syllogistics and modern formal logics since the second half of the 19th century appeared too great. Some authors simply divided the history of their discipline into two areas: the old and the new logic.

In recent decades, however, researchers have repeatedly shown that there are old logics with enormous potential: in Arabic and islamic logic the consequence relations show inspiring alternatives to modern approaches. Today, Indian logic is often associated with paraconsistency and dialetheism and in the field of natural language processing, medieval logicians are increasingly used to circumvent the artificiality of algebraic logic. In modal logic, Aristotelian and scholastic logics are again increasingly discussed. Numerous modern systems in the field of visual reasoning are based on the Aristotelian square of opposition, on the arbor porphyriana or on Euler's logic diagrams. And in early modern period new propositional calculi and extended syllogistics are discovered frequently, which pose challenges to interpretation.

The fact that in recent decades many modern innovations have been able to draw on ideas in logic that are actually 'old' led us to organise a two-day workshop in early December 2019 as part of *Creativity: 1st*

World Congress of the Brazilian Academy of Philosophy in Honour of Newton da Costa 90th Birthday. This workshop bore the same title as the present volume, i.e. *History of Logic and its Modern Interpretation.* Numerous researchers, especially from South and North America and from Europe, gathered in Rio to lecture on various logical topics in the history of logic and to present new approaches to interpretation. During the personal exchange of ideas at the workshop, we also discussed which publishing companies were available to modern researchers for the broad history of logic. We noticed that there were excellent journals such as *History and Philosophy of Logic* that published stand-alone papers or reviews and that many journals that only published technical papers on logic 50 or 30 years ago were now also opening up to traditional topics from the history of logic. Nevertheless, in our opinion, an established book series on the history of logic was lacking.

Our idea was therefore to found such a publication series, initially focusing on publications from the English and German language areas, with the option – if this series is successful and the editorial board expands – to include other books from other national languages.

We approached College Publications a few months after returning from Rio. Dov Gabbay and Jane Spurr, to whom we are deeply indebted, supported our idea and we soon gathered an editorial board of 22 researchers. We wanted them to be able to review both English and German language publications, to cover all parts of the history of logic, i.e. with both continental emphasis and knowledge of the varied historical periods, and also to take into account the different subject areas of logic, i.e. from philosophy to cultural studies, theology, linguistics, mathematics and computer science. To make it clear that we wanted to focus not only on the history of modern logic, but on the entire history of logic, we decided on a Latin title for the book series, *Historia Logicae.* Historia Logicae aims to publish high-quality monographs, dissertations, textbooks, proceedings, and anthologies written in German or English. All contributions should have a clear reference to the history of logic, both history and logic being understood in the broadest sense, i.e. without restriction to any period, any region, any logic or any discipline related to logic: The time span ranges from pre-Aristotelian approaches to debates reaching to the present; regionally, the contributions can extend to all continents, cultures, and traditions; and besides the various logics, contributions to logic-related disciplines such as mathematics, rhetoric, linguistics, ontology, artificial intelligence, computer science, etc., are also welcome if they contain a recognizable historical interest.

As we later noticed, the title *Historia logicae* was not new either: in 1721, Johann Georg Walch had already published in Leipzig a text under this title, and as we quickly discovered, this could be interpreted as a good omen. Walch was not the first to write a history of logic in the early modern period. He himself lists numerous predecessors in his work, such as Bartholomäus Keckermann's *Praecognitorum logicorum* (1598) or Pierre Gassendi's *De origine et varietate logicae* (1658), and critically draws on them in his own account of the history. Due to the thoroughness of Walch's approach and due to the fact that he confronted many historical details with the modern views of his time, he can be considered a thoroughly good namesake for our book series. We then published a CfP in 2020 to have contributions from the Rio workshop, but also other papers, reviewed by the editorial board. The demand was enormous and after an intensive review process, which was particularly complicated by the Corona pandemic from 2020 onwards, we arrived at the twelve contributions that are collected here in this volume. The aim of the review process was not to select one or more papers from each subject area or from each era, but to identify the best papers that can show how to interpret the old, or at least time-barred, topics from logic by modern means. We, therefore, had no claim to completeness as far as the history of logic was concerned. For this reason, the anthology presented here does not read like a continuous history of logic (e.g. in the vein of Walch). Rather, it can be understood as an insight that shows which topics from the history of logic are being researched today and what possibilities of modern interpretation there are. It can also be understood as a documentation of the congress in Rio, since several of the contributors to this volume also presented their topics at the workshop.

The volume as a whole is divided into two parts and the essays have been arranged – as far as this was possible – according to the historical course of the texts, which are interpreted in a modern way: The first part deals with ancient and medieval logic. Plato, Aristotle, Seneca, Porphyry,

Proclus, Boethius are discussed here as well as Buridan or Kilwardby. Thematically, the relationship between mathematics and logic, natural deduction, intuitionism, graph theory, categorical judgements, theories of truth and modal logic are addressed. The second part then covers the early modern and modern periods: Here, texts and topics by and on Caramuel, Kant and Kantians, Husserl and even Jaśkowski are discussed. Thematically, papers on opposition relations, the definition of formal logic, many-sorted logic, model theory and paraconsistent logic are covered. The abundance of topics and texts should therefore be further proof for the reader that a modern interpretation of the history of logic makes perfect sense. We would like to thank the reviewers and sub-reviewers (Lorenz Demey, Ulf Hlobil) of this volume, all researchers who submitted papers, the participants and discussion partners at the Rio meeting, the team at College Publications and especially Alfred Olszok, who did valuable work in editing the papers.

September 2022   Jens Lemanski (Münster) & Ingolf Max (Leipzig)

Part I

Ancient and medieval logic

Self-Reference and Type Distinctions in Greek Philosophy and Mathematics

IOANNIS M. VANDOULAKIS
Hellenic Open University, Greece and FernUniversität in Hagen
`i.vandoulakis@gmail.com`

Abstract. In this paper, we examine a fundamental problem that appears in Greek philosophy: the paradoxes of self-reference of the type of "Third Man" that appears first in Plato's *Parmenides*, and further discussed in Aristotle, and the Peripatetic commentators, and Proclus. We show that the various versions are analysed using different language, which reflects different understandings by Plato and the Platonists, such as Proclus, on the one hand, and the Peripatetics (Aristotle, Alexander, Eudemus), on the other hand. We show that the Peripatetic commentators do not focus on Plato's solution but primarily on the statement of the paradox. On the contrary, Proclus seems to be convinced that Plato suggests a sound solution to the paradox by the definition of the predicate of *similarity* (homogeneity) that demarcates two types of homogeneous entities – the *eide* and the participants in them, in a way that their confusion would be inadmissible.

We claim that Plato's solution follows a sound line of reasoning that is formalizable in a language of Frege-Russell type; hence there exists a model in which Plato's reasoning is valid. Furthermore, we notice that Plato's definition of the second-order predicate of similarity is attained by resorting to first-order entities. In this sense, Plato's definition is comparable to Eudoxus' definition of *ratio*, which is also attained by resorting to first-order objects. Consequently, Plato seems to follow a logical practice established by the mathematicians of the 5th century, notably Eudoxus, in his solution to the paradox.

1 Introduction

The well-known Third Man Argument in Plato's *Parmenides* 132a-133b has received much attention in the scholarly literature and undergone meticulous analysis by logicians, philosophers, and classicists, especially after Vlastos' seminal paper "The 'Third Man' Argument in the *Parmenides*" [31] in which it is interpreted for the first time as a paradox of self-reference.[1]

Literarily speaking, in Plato's *Parmenides* appears an argument that could be called "the Third Large" argument, because the predicate "is large" is used to generate an *aporia* (difficulty). The name "Third Man" appears in the surviving texts of Aristotle, i.e., *Metaphysics*; *Sophistic Refutations*, and the texts of Aristotelian commentators, notably Eudemus and Alexander of Aphrodisias. It is commonly considered that these arguments are logically the same and differ mainly in the choice of the concrete predicate.

In this paper, we will examine these versions of arguments and their logical structure and focus on certain semantic aspects. As we will see, the language used in the Platonic and the Peripatetic versions are different and this reveals different understandings by Plato and Neo-Platonic philosopher Proclus, on the one hand, and the Peripatetic philosophers, on the other hand. Their view and evaluation of the argument are different.

In the Peripatetic tradition, the "Third Man" is interpreted in terms of predication and qualified as a fallacy. The Peripatetic philosophers focus primarily on the statement of the argument and evoke implicit premises on which, in their view, it is grounded; they neglect Plato's

---

[1]Summaries of interpretations of the Third Man Argument and relevant bibliographies are to be found in [22]; [6, 32 n. 18] and [25]. A critical survey of more recent approaches can be found in the Appendix in [24].

discussion and suggested solution using the concept of "similarity" in *Parmenides*.

On the contrary, in the Neo-Platonic tradition, the name "Third Man" is not used. The "Third Large" argument is stated using a metaphor language specific to Plato, interpreted in terms of modes of participation and qualified as a major *aporia* which nevertheless is surpassable by following the line of Plato's reasoning. Special attention is paid by Proclus to the concept of "similarity" and how it applies to the particulars and the Forms, which is the key to Plato's solution of the paradoxicality.

2 The 'Third Large' Paradox in Plato's *Parmenides*.

2.1 Statement of the argument

First, we will expose concisely the "Third Large" aporia as it appears in Plato's *Parmenides*, following our reading of the text as justified in [30].

The argument begins by stating the assumption that when some *idea* (of being large) is viewed over all the things of a plurality, then one can conclude that the *eidos* (the Form of the Large, or the Largeness) F_0 is *one* (in which participate all the things of the plurality).

Then, one can view 'in the same way' over all these large things and the Form of the Large F_0 the same *idea* (of being large) by virtue of which they all appear to be large. Thus, another *eidos* (Form of the Large) F_1 will appear generated by the Large F_0 and the participants of it, i.e., in the Form F_1 participate all the things of the plurality and the Form F_0 itself.

By applying the same line of reasoning, by viewing the same *idea* (of being large) over them all (the things, the Form F_0 and the Form F_1), another *eidos* (Form of the Large) F_2 is generated in which participate all the things of the plurality, the Form F_0, and the Form F_1.

This procedure can be continued *ad infinitum*, giving rise to an infinite sequence of new expanding Forms. This contradicts the initial Uniqueness Assumption that every plurality of things corresponds to one Form (*eidos*).

2.2 Plato's language of the Third Large Argument

We will now highlight certain aspects of the text and the vocabulary used which are crucial for the logical structure of the argument and its understanding.

A key distinction for the logical understanding of the argument is between the concepts of *eidos* (εἶδος) and *idea* (ἰδέα). In scholarly literature, the corresponding terms 'Form' and 'Idea' are commonly used interchangeably to render a whole cluster of Greek terms, such as εἶδος, ἰδέα, οὐσία, γένος, which are close in meaning but not identical in any given context. Insofar as Plato has not provided us with a systematic treatment of his theory of Forms, nowhere in his *Dialogues* explains how his terminology is used and must be understood or how various terms differ from each other in different contexts. Both these terms, *eidos* (εἶδος) and *idea* (ἰδέα), are etymologically associated with the metaphor of "viewing", which plays a crucial role in the statement of the argument. All three terms derive from the common root ϝιδ (see the entry εἶδος in [16]). An *idea* is a common feature or character that is viewed to stand over the participants of a plurality, whereas *eidos* indicates a rather completed act of "viewing over," that is the visualised plurality of particulars.

The distinction between εἶδος and ἰδέα has become a subject of systematic inquiry by Russian philosopher Aleksey F. Losev (1893 – 1988) in 1930 [17] and Dutch classical philologist Peter Brommer (1892 – 1982) in [5].[2] Despite the divergence of their approaches and interpretations, an *eidos* is an entity generated out of constituents and divisible into parts, whereas an *idea* is an entity of atomic (partless, indivisible) nature. Moreover, an *idea* appears in many contexts to have a unifying function that "embraces" or "comprehends" a plurality of things or "brings them to" one *idea*.[3] [*Phaedrus*, 265d, 273e] [*Laws* 12.965c] [*Sophist* 253d].

The Third Large argument is obtained by the repeated usage of the

---

[2] See also Losev's review of Brommer's book in [18] and Rimondi's study on Losev's terminological investigations on *eidos* and *idea* in [26].

[3] For a detailed analysis of these terms, see [30, pp. 6 - 11].

Self-Reference and Type Distinctions

metaphor of "viewing an *idea* over all things" (μία τις ... ἰδέα ἡ αὐτὴ εἶναι ἐπὶ πάντα ἰδόντι). The *idea* is viewed over all the particulars, at the first step of the argument, the particulars and the *eidos* F_0, at the second step of the argument, and so on. Throughout all these steps, the *idea* (of being large) remains unchanged and applies to the particulars, to the *eidos* F_0, to the *eidos* F_1, and so on, 'in the same way' (ὡσαύτως), i.e., as if the *eide* $F_0, F_1, \ldots$ were single particulars. Thus, the notion of the *idea* names a feature or property common to all things (ἡ αὐτὴ εἶναι ἐπὶ πάντα) to define the *eide*, which take their names from this *idea*. It is associated with the metaphor of "viewing over" things in contexts where it appears as a unifying principle of a plurality of particulars, whereas the *eide* are related with the things by the concept of "participation" (μέθεξις, μετέχειν), which is a fundamental concept of Plato's ontology.

The word "all" (πάντα) plays the logical function of a quantificational word. "All" focuses on the plurality of large things, which are turned out within the scope of the act of "viewing over" them. This act is supposed to take place in mind (τῇ ψυχῇ ἐπὶ πάντα ἴδῃς) [11]. In the subsequent steps, the scope of the act of "viewing over" all particulars is extended over the newly generated *eide* $F_0, F_1, \ldots$ taken as if they were particulars. In this way, the *idea* 'large' is taken to refer not only to the initial particulars but also to the *eide* $F_0, F_1, \ldots$ (the Forms of the Large), i.e., the *eide* $F_0, F_1, \ldots$ are 'large' or, as is commonly concisely stated "Largeness is large" (self-reference).

Special attention should be paid to the pronoun ᾧ, in the phrase ᾧ ταῦτα πάντα μεγάλα φαίνεσθαι. This pronoun occurs in the dative case and is usually rendered as "by (virtue of)" or "because of." However, in *Parmenides*, there is no subject in the dative case to which ᾧ could refer. Thus, ᾧ is an elliptical phrase. An examination of this phrase in other Platonic passages where is used shows that ᾧ refers to an *idea* [30, pp. 13 - 14]. For instance, in *Euthyphro*, the word μίᾳ ἰδέᾳ is found in the dative case. Thus, we can conclude that pronoun ᾧ refers το μίᾳ ἰδέᾳ.

> The *eidos*, by which (ᾧ) all holy acts are holy; for you said that all unholy acts were unholy and all holy ones holy by one *idea* (μίᾳ ἰδέᾳ).[*Euthyphro 6d*]

Consequently, the new *eide* in Plato's Third Large Argument are generated at each step because the corresponding entities (particulars and the *eide*) are viewed to be large by virtue of an *idea*.

A crucial point in the generation of the infinite sequence of new *eide* is that the act of viewing over the entities inferior to an *idea* at each step is implemented "in the same way" (ὡσαύτως) [11], i.e., each newly generated *eidos* is treated as an individual in terms of logical behaviour. Thus, the *idea* of "large" (μέγα) plays a double role. On the one hand, it is used as a general name or "class-as-many" to generate new *eide*, and, on the other hand, it is used as "class-as-one" (αὐτό τὸ μέγα), i.e., as a singular entity "alongside" (παρά) the Form of the Large and the participants in it. In other words, the *eidos* generated at each step is subsequently treated as if it were *on the same rank* with the particulars that participate in it. This allows the extension of the scope of the quantificational word "all" over both every new *eidos* and the things participating in it. The term "large" (*idea*) is considered as stating something about *all* the entities, including the new entities that are been formed at every step. Hence, the idea "large" is applied *in the same way* to both particulars and the *eidos* of the Large, which is understood as behaving like an individual thing. This explains the logical mechanism of the Third Large Argument.

Consequently, the inconsistency for Plato's theory of Forms becomes clear:

> "each of the *eide* will no longer be one for you, but infinite in number (ἄπειρα τὸ πλῆθος)".[*Parm.* 132b]

Thus, the Uniqueness Assumption ceases to wind down. Consequently, it is not the infinite regress itself that is considered as logically incoherent, but rather that self-participation beclouds the Uniqueness Assumption.

2.3 Plato's solution of the Third Large Paradox

To find a way out of the paradoxical situation, Socrates translates the question from a question over Forms into a question over some new mind-dependent entities, called *noemata* (νοήματα, "thoughts"). The word *noema* is rather rare in Plato, whereas it is a key concept in the historical Parmenides' poem *On Nature*. Thus, Allen rightly concludes that Socrates' suggestion echoes the historical Parmenides [2, pp. 148 - 149]. It is not clear how this is meant to block the paradox. The subsequent line of reasoning is rather obscure. Nevertheless, the suggestion is rejected although it is not clear on what grounds. Hence, we can conclude that the *eide* must not be understood as mind-dependent entities. This opens the way to the next approach initiated again by Socrates.

The *eidos* is defined to be a *paradigm* (παράδειγμα), which expresses the form of instances (ὁμοιώματα, 'copies,' 'resemblances') of the *eidos*, "found" in nature. Here, it is implicitly assumed that the 'copies' are instances of a predicable entity, that is of an *idea*. In this way, participation in an *eidos* can be considered equivalent to instantiation (εἰκασθῆναι - 'imaging') of the *eidos*, as stated in the text. [*Parm.* 132b] Nevertheless, these new assumptions do not prevent yet the generation of new *eide* by using self-reference, since intensionally considered the paradigm can become an instance of a further paradigm (this could be called *self-instantiation* paradox).

The next step in the argument begins with fixing an arbitrary particular thing (τι),[4] which is presumably an instance of an *eidos*. Further, an *eidos* is compared with a fixed instance of it (τι ... ἔοικεν τῷ εἴδει). In terms of logic, a two-place predicate is introduced:

'_ is similar to ...' (ὅμοιον εἶναι),

where "_" stands for an eidos and "..." for a fixed instance of it, and the following questions are posed:

- Can we conclude that an *eidos* is similar to an instance of it (τῷ εἰκασθέντι) on the ground that the latter is an 'instantiation' of the *eidos* (καθ' ὅσον αὐτῷ ἀφωμοιώθη)?

- Is there any way in which an *eidos* is not similar to an instance of it?

The immediate answer given in the text is that there is no way for an *eidos*, be not similar to the imaged [thing]. From this follows that an *eidos* must be similar to an instance (participant) of it.

---

[4]Generally, there is no word for "thing" in the Greek text of Plato's *Parmenides*. In the beginning of the argument, Plato uses the phrase πόλλ' ἄττα, which is commonly rendered as "a plurality of things." However, in this step he fixes a particular thing by naming or indicating it (τι).

Nevertheless, this conclusion does not prevent self-instantiation yet. Because if an *eidos* is similar to an instance of it, then another *eidos* may appear in which the first *eidos* and its instance both participate. This argument is commonly called *resemblance regress*. In this generation of the infinite sequence of *eide*, the property "being similar" plays the role of an *idea* viewed over the *eide* and their instances.
However, this possibility is excluded by the following statement:

"It is not possible for some [thing] to be similar to an *eidos*,
or the *eidos* [to be similar] to any other [thing]".[*Parm*. 132e]

The meaning of this proposition is usually overlooked. This statement presupposes an understanding of similarity different from that described above that leads to the resemblance regress. The new meaning of similarity is introduced in the preceding proposition:

"It is absolutely necessary for the like [thing] to participate
in one and the same *eidos* as what it is like".[*Parm*. 132e]

In other words, two things (one of which is a fixed instance of an *eidos* - the like [thing]) are similar to each other if and only if they "participate in one and the same *eidos*", namely in the *eidos* in which the fixed instance participates. In other words, if x is a fixed instance of F and z any other thing, then z is defined to be similar to x if and only if z participates in the same *eidos* F of which x is a fixed instance.

In this way, z ranges over what can be called today the *domain* of the *eidos*, i.e., the collection formed by the participants of the *eidos* of which x is a fixed instance or, equivalently, by the instances of the *eidos* of which x is a fixed instance. This 'domain' obviously consists of *homogeneous* things, i.e., 'similar' to each other. This restricts the formation of *eide* insofar now their 'domain' should be taken into account. As a result, *eide* and things are not homogeneous entities.

This defines a demarcation line between two kinds of homogeneous entities: the level of particulars and the level of the *eide*. The confusion between non-homogeneous entities is not allowed. The lower level (the 'domain') of the particular things is defined by the possibility to establish the relation of similarity (dissimilarity) among them, while the relation of the particulars of the lower level to the entities of the higher level, i.e., to the Forms (*eide*), is defined by participation. This stratification of entities and the corresponding restriction on the formation of *arbitrary* Forms, i.e., by confusing non-homogeneous entities, remove the Third Large paradox from Plato's theory of Forms. Otherwise, i.e., if non-homogeneous entities are taken to be homogeneous,

> "alongside the eidos, another eidos will always appear, and if the latter is similar to the first, another one again and will never stop a new eidos to be generated every time, if the eidos is made similar to a participant of it.[*Parm*. 132e - 133a]

The last conclusions in Plato's argument follow now naturally. Under this definition of the concept of similarity, it is clear that the relation between particulars and Forms ('participation') cannot be defined by any relation, which holds among particulars i.e., 'similarity':

> "the other [things] do not partake of the *eide* by means of similarity."[*Parm*. 133a]

The whole discussion closes with the ascertainment that the difficulty (*aporia*) is due to the establishment of the *eide* as if they were in themselves. [*Parm*. 133a] The text suggests that at this point the problem is considered resolved.

3 The Peripatetic view of the 'Third Man' Paradox

Although the Third Large argument might have been considered resolved in Plato's Academy, it seems that this was not certain in the Peripatetic school. Aristotle and later commentators show a critical attitude towards an argument called "Third Man" in the texts of the Peripatetics.

In the surviving texts of Aristotle and the Peripatetic commentators, we do not find the Third *Large* Paradox exactly as Plato has formulated and discussed in *Parmenides*. There are only scattered references to an argument that Aristotle calls the "Third *Man*".[*Met.* 84.23-85.3, 93.1-7, 990b 17=1079a 13, 1039a 2, 1059b 8; *Soph. El.* 178b 36] These references are not very illuminative, because they only mention the argument; neither formulate nor analyse it in detail.

The texts of the Peripatetic philosophers focus primarily on the statement of the argument which is reformulated by using certain implicit premises that they consider as required for its deduction. Neither the specific Platonic metaphorical language is used, nor the concept of similarity is ever discussed. The latter can be possibly explained by Aristotle's rejection of Forms as existing in themselves; hence, the question of similarity or dissimilarity between the Forms and the particulars has no sense for the Peripatetics. Thereby, it seems that Plato's solution of the paradox by appealing to the concept of similarity and the establishment of a demarcation line between two domains of homogeneous entities, the particular things and the Forms, was not acceptable by the Peripatetic commentators.

3.1 Aristotle's version of the Third Man Argument.

Alexander ascribes the following formulation of the Third Man directly to Aristotle [1, pt. 85.11 - 12]:

> If what is predicated (τὸ κατηγορούμενον) of some plurality of things, i.e., 'man,' is also another thing, i.e., the man-itself, besides (παρά) the things of which it is truly predicated (ἀληθῶς κατηγορεῖται), being separated (κεχωρισμένον) from them ... there will be a third man. In the same way, there will be also a fourth, fifth man, and so on to the infinite. [7, p. 19]

Plato's metaphor of the mind's "viewing of an *idea* standing over all particular things" is substituted by the Aristotelian concept of predication. Thus, 'man' plays the role of a general term truly predicated of a plurality of things, which generates a new entity, a universal in Aristotle's sense. This is the Aristotelian version of what is commonly called the *One over Many Principle*.

However, there are several aspects in this argument that need further examination. The newly generated entity

1) Is *different* (ἄλλο) from each thing of the plurality.

2) Lies *besides* (παρά), i.e., on the same rank with the things of which is predicated.

3) Is *separated* (κεχωρισμένον) from the things of which is predicated.

4) *Subsists on its own* (κατ' ἰδίαν ὑφεστώς), i.e., it has an independent existence.

The last notion of form as that which is capable of independent existence is associated with Aristotle's concept of *substance* (in one of the senses he uses it), which is not mentioned in the text. Aristotle uses the notion of substance in this sense to emphasize the distinction between substances and their qualities or relations.

Are these features relevant to Plato's formulation of the Third Large in *Parmenides*?

The first feature, i.e., that the generated entity is different from each thing of the plurality is found in Plato's formulation of the Third Large, i.e., in the phrase:

> "another (ἄλλο) *eidos* of Largeness will appear ...".[*Parm.* 132a]

The second feature, i.e., that the generated entity lies *on the same rank* (παρά) with the things of which is predicated is relevant to the Platonic text since, in Plato's Third Large Argument, the new *eidos* is

> "generated alongside (παρά) the Largeness and the participants of it".[*Parm.* 132a]

Even the words used by Plato and Aristotle coincide, i.e., ἄλλο (another) and παρά (besides, alongside).

The other two terms in 3) and 4) are not used in Plato's text, but the meaning of these statements seems to appear in the concluding sentence of the Third Large Argument:

> "Do you see now, Socrates, how great the difficulty (*aporia*) is, if someone establishes the *eide* as if they were in themselves? (ἐάν τις ὡς εἴδη ὄντα αὐτά καθ' αὑτά διορίζηται;)[*Parm.* 133a]

Aristotle was convinced that Plato separated Forms from the things and that separation is responsible for many difficulties in Plato's theory of Forms, including the 'Third Man'. However, the separation is not a necessary premise for the infinite regress and is not even mentioned by Plato. Moreover, Aristotle emphasises in this version the independent subsistence of Forms, a point that in Plato appears only in the last sentence quoted above.

Consequently, we can conclude that the Aristotelian language used is relevant to Plato's Third Large Argument, except for the concept of predication.

On the other hand, several aspects characteristic of the Platonic language and understanding are missing in the Aristotelian version.

(i) There is no distinction between *eidos* and *idea*.

(ii) There is no use of the term "participation" and "participant" which is characteristic of the Platonic theory of Forms.

(iii) There is no use of the metaphor of "viewing over" things.

(iv) There is no use of the specific word ᾧ.

(v) There is no use of the quantificational word "all" (πάντα), which is essential in the generation of the Third Large argument. However, in the Aristotelian version the term "truly predicated" is used, which can be considered as a logically equivalent concept.

(vi) There is no explicit formulation of the Uniqueness Assumption, i.e., the thesis that "each *eidos* is one." This thesis is essential in the argument because its violation is viewed as paradoxical, as suggested by Plato's words "each of the *eide* will no longer be one for you, but infinite in number."

As Plato relies on the concept of self-participation for the deduction of the Third Large argument, so Aristotle relies on the concept of self-predication, which is not explicitly formulated, although it is necessary for the generation of the argument. Besides, Aristotle had to require also that the common name 'man' be predicated *in the same way* of both the particulars and of the form. This is explicitly expressed in Plato, by the word ὡσαύτως ("in the same manner").

Finally, there is no reference or comment in the Aristotelian version to the Platonic concept of similarity and Plato's solution using this concept.

3.2 The Eudemean version of the Third Man Argument.

Alexander ascribes another version of the Third Man Argument to Eudemus, which differs significantly in terminology from both Plato's Third Large and Aristotle's version of the Third Man Argument. The argument proceeds from the assumption that

> "The things that are predicated in common of substances (τῶν οὐσιῶν) are *properly* (κυρίως) [substances] and are forms".[5]

The use of the term 'substances' seems to be used in the sense of Aristotle's concept of 'secondary substances' in the *Categories*, i.e., for entities predicable of a subject. "For instance, 'man' is predicated of the individual man" [Aristotle *Categories* 2a 21–22], as when we say, "Socrates is a man." Aristotle seems to have the idea that essences (οὐσίαι) are substances, and the more qualities they comprise, the more substantial they truly are:

> Of the secondary substances, the species is more truly (μᾶλλον) substance than the genus, being more nearly related to primary substance. [27, pt. 2b 7]

According to Aristotle, substances can exist on their own. Hence, we can conclude that the use of the term 'substance' in the text implicitly assumes that the form of man has independent subsistence, as stated explicitly in Aristotle's version. Aristotle explains that primary substances are called *properly* substances (κυριώτατα οὐσίαι λέγονται),

> "because they underlie and are the subjects of everything else". [27, pt. 3a 1–3]

---
[5]my translation; cf. [7, p. 19]

Consequently, the Eudemean version of the Third Man is formulated in a language of predication derived from Aristotle's theory of categories. This concept of predication is explained by recalling Plato's definition of similarity (ὅμοιον εἶναι), which is reformulated in the following way:

> "things that are similar to one another are similar to one another by sharing (μετουσία) in some same thing, which is *properly* (κυρίως) this [i.e., properly substance]; and this is the form". [6]

Note that the term μετουσία is used here in the dative case, which corresponds to Plato's use of the term ᾧ (μίᾳ ἰδέᾳ).

In this way, the starting point of the argument is the 'domain' of homogeneous things which participate in one and the same *eidos* (in Plato's terminology).[*Parm.* 132e]

Then the argument proceeds from the assumption that if what is predicated in common of things is *different* (ἄλλο) from any of those things of which it is predicated, then a new entity will appear besides it (παρ' ἐκεῖνο). In this exposition, the root of the problem seems to be that the new entity is *different* from the things. However, this is not sufficient to deduce the Third Man Argument. Self-predication is necessary, which is not explicitly formulated. If we follow the Eudemean line of reasoning, we must assume that the new entity should be taken to be similar to the things, or, in Plato's words,

> "the *eidos* is *made* similar to a participant of it (τὸ εἶδος τῷ ἑαυτοῦ μετέχοντι ὅμοιον γίγνηται)"[*Parm.* 132e - 133a]

This assumption would then generate the so-called Resemblance Regress. However, the Eudemean version contains no reference to any kind of regress. The argument stops at the second step, i.e., at the generation of the 'third man,' without indication that this line of reasoning can be repeated *ad infinitum*. The Uniqueness Assumption, i.e., the thesis that "each form is one" appears in passing only at the end of the argument

> "then there will be a third man besides the particular (such as Socrates or Plato) and besides the idea, *which is also one in number*". [7, p. 19]

---

[6]my translation; cf. [7, p. 19]

However, this statement does not make clear that the paradoxicality of the Third Man Argument consists in the violation of the Uniqueness Assumption.

3.3 Alexander's version.

Alexander's version of the 'Third Man' [1, pt. 83.34 - 85.12] seems to be a mélange of the Aristotelian and the Eudemean versions.

In comparison to the Eudemean version, there is neither reference to properly (κυρίως) substances, nor to any substances, in the sense of Aristotle's *Categories*. However, the similarity is explained by using the term μετουσία, in the same way as in the Eudemean version. However, in Alexander's version the assumption that "men and the forms [of men] are similar" is stated explicitly as the cause of the problem in the argument. Nevertheless, Plato's cautious language that the *eidos* is *made* similar to a participant of it is absent in this version as well.

In comparison to the Eudemean version, the Uniqueness Assumption is not stated, even in an ambiguous form. On the contrary, it is explicitly stated that the process of generation of forms can be continued *ad infinitum*.

It is noteworthy that in Alexander's version there is, for the first time, an explicit indication that the paradox does not depend on the concrete predicate ('man'). Specifically, it is stated that

> "a similar multiplication will be suffered *by each of the other things of which they say there are ideas* [be they Aristotelian forms or Platonic *eide*]". [7, p. 19]

This text gives us grounds to assume that Plato's Third Large and the Aristotelian Third Man Arguments were perceived as logically equivalent. They both refer to the same logical argument, although their understanding and interpretations differ depending on the conceptual and semantic framework used in the Platonic and the Aristotelian traditions respectively.

It seems that the Peripatetics were also aware that the paradoxicality concerns not only Plato's theory of Forms (*eide*) but also the Aristotelian theory of forms and the theory of categories. Otherwise, there would have been no reason to discuss it in their terms, i.e., in terms of Aristotle's theory of substances. They would have presented it just as an argument refuting Plato's theory of Forms, i.e., as a fatal argument refuting a rival theory. But we find no such presentation. How then the Aristotelians solved this problem for the Aristotelian theories of forms and categories?

4 Self-reference and the Aristotelian theory of categories.

Let us consider why the Third Man argument is relevant to Aristotle's theory of categories. A fundamental distinction in Aristotle's theory of categories is between the primary and secondary substances.

> "A *substance*—that which is called a substance most strictly (κυριώτατα), primarily (πρώτως), and most of all—is that which is neither said of a subject nor in a subject, e.g. the individual man or the individual horse. The species in which the things primarily called substances are, are called *secondary substances*, as also are the genera of these species. For example, the individual man belongs in a species, man, and animal is a genus of the species; so these—both man and animal—are called secondary substances." [4, *Categories* V 2a 11 - 19]

The criterion for this distinction into primary and secondary substances seems to be the following:

(a) being subject but never predicate,

(b) not found in a subject.

Accordingly, the primary substances satisfy both conditions, whereas the secondary substances satisfy the second condition but not the first. Consequently, primary substances are individuals, while secondary substances are species and their genera, which are predicable of a subject.

> "It is clear from what has been said that if something is said of a subject both its name and its definition (τὸν λόγον) are

necessarily predicated of the subject." [4, *Categories* V 2a 19 - 20]

Therefore, the primary substance is the ultimate subject of predication, which assume the form:

- The (primary substance) is predicated of the name of the species;
- The (primary substance) is predicated of the definition of the species.

On the contrary, for things that are found in a subject, neither their names nor their definitions can, in most cases, predicate of that subject.

> "But as for things which are in a subject, in most cases neither the name nor the definition is predicated of the subject. In some cases there is nothing to prevent the name from being predicated of the subject, but it is impossible for the definition to be predicated." [4, *Categories* V 2a 27 - 30]

When something (a genus or a species) is predicated of another thing as of a subject, everything said of the predicate is also said of the subject [4, *Categories* III 9 - 15]. Species is related to the genus as a subject is related to a predicate. The genera are predicated of species, but conversely, the species are never predicated of the genera.

> But as the primary substances stand to the other things, so the species stands to the genus: the species is a subject for the genus (for the genera are predicated of the species but the species *are not predicated reciprocally* of the genera). (τὰ μὲν γὰρ γένη κατὰ τῶν εἰδῶν κατηγορεῖται, τὰ δὲ εἴδη κατὰ τῶν γενῶν οὐκ ἀντιστρέφει.) – our emphasis. [4, *Categories* V 2b 19 - 21]

Consequently, the following restrictions upon predication are imposed by Aristotle:

(i) nothing is predicated of individuals,

(ii) the genera are not predicated of species.

Aristotle allows

a) the species to be predicated of individuals;

b) the genus to be predicated of the species and individuals.

c) everything said of the predicate to be also said of the subject.

Under these conditions, the Third Man argument cannot appear in Aristotle's theory of categories. The latter is a logically consistent theory, free from paradoxes of self-reference. Aristotle constructs his theory taking into account the problems that self-predication might cause.

5 Proclus' commentary of the 'Third Large' Paradox.

Although Proclus offers the most elaborate and lengthy commentary of the Third Large argument, it has not attracted attention in the scholarly literature, except for the work of Gerson P. Lloyd in [19]. Proclus follows scrupulously Plato's line of reasoning and advances consistently the metaphysical problems associated with the argument. He is convinced that the argument is not a real thread for Plato's theory of Forms and presents it just as a logical exercise in the mode of participation.

Proclus describes Plato's Third Large argument by using a scheme of ascension from the sensible things upwards to their Forms. In this process, he distinguishes three levels of Being, namely

a) the level of the sensible things, termed as "the participants" (τὰ μετέχοντα).

b) the level of the reason principles in Nature (φυσικοὶ λόγοι) which act to hold the species of sensible things together.

c) the level of Forms, termed as "the participated entities" (τὰ μετεχόμενα).

Thus, he describes the argument as follows:

"When we see a multiplicity of large things and perceive a single *idea* pervading them all (ἐπὶ πάντα ταῦτα μίαν ἰδέαν διατείνουσαν), we hold that there is one Largeness common to all the instances of largeness in the individual things (τοῦ ἐν τοῖς καθέκαστα κοινόν). That the argument is about the natural species and the transition to it from sensible things he [Plato] makes clear by the insertion of such terms as *thinking* (οἴεσθαι), *it seems to you* (δόξῃ), *you consider* (δοκεῖ), none of these is a term that can be used with respect to objects of scientific knowledge, but only about the realm of nature."[Procl. *in Parm.* 879; Morrow, 241]

It is noteworthy that Proclus pays special attention to Plato's language and the metaphors used by such terms as *thinking* (οἴεσθαι), *it seems to you* (δόξῃ), *you consider* (δοκεῖ), which he associates not with science but with the realm of nature. He also keeps the Platonic term *idea* to denote the specific meaning of the unifying principle over the sensible particulars. An *idea* is viewed as "pervading over all sensible particulars."

Moreover, Proclus explicitly recognises that the predicate (the large) chosen does not affect the logical structure of the argument. He explains that by using the predicate "man" one generates the Third Man argument[Procl. *in Parm.* 879; Morrow, 241] However, he does not use the specific language we know from the Peripatetic commentators but his aforementioned scheme of the three levels of Being.

In using this scheme, he clarifies that the ascension from bodies to natures implies that the "communion" (κοινωνεῖν) between the one Form and its many instances is not merely nominal, because of the *common name* (κοινὸν ὄνομα) or *synonymy* (συνώνυμον) between the one and the plurality, but

> "the common element in the many instances is that of being derived from and having reference to a single source. For what the one Form is primarily, the many grouped under it are derivatively."[Procl. *in Parm.* 880; Morrow, 242]

He further clarifies that

> "although community belongs to things coordinate (ὁμοταγῶν) in rank, yet it is not coordinate with the things that have community"[Procl. *in Parm.* 880; Morrow, 242]

In other words, sensible things that share a common feature belong to one level but their common feature does not belong to the same level. The confusion of these levels, i.e., the level of the participants and the level of their common property, is responsible for the infinite regress:

> "If there is anything in common between Forms and the things which participate in them, the infinite regress will emerge as between those things participating in the common property and the common property itself."[Procl. *in Parm.* 882; Morrow, 243 - 244]

Thus, Proclus seems to be aware of the root of the Third Large paradox. In his view, the fallacious ascension is caused by the incorrect mode of participation when understood as a communion between the participated entities and their participants.[Procl. *in Parm.* 885; Morrow, 246] He explains that communion cannot be the causal principle of participation, namely, he clarifies that

> "for, if there were some property in common between the participant and what is participated in, we will have to make a transition again from these to something which would be the cause of the communion of both of them, and so one would proceed to infinity; for in the case of any things which have a property in common, there must be something prior to them which is the cause of their community."[Procl. *in Parm.* 886; Morrow, 247]

Thus, in Proclus's view, the ascension for sensible things to Forms in Plato's *Parmenides* 132a-b is not correct, because the reason-principles of nature that function as principles of unity in nature that generate all forms of sensible things are taken to be on the same level of being as the sensible things.[Procl. *in Parm.* 887; Morrow, 248]

Proclus seems convinced that Plato is aware that the argument is invalid. He shows that by pointing to Plato's use of the word "yours" (σοι) in the phrase "and so each of *your* Forms will no longer be one" (καὶ οὐκέτι δὴ ἓν ἕκαστον σοι τῶν εἰδῶν ἔσται). This locution implies, in Proclus's view, that the infinite regress can take place only in individual minds, not in the realm of the Forms.

In connection with Plato's interpretations of Forms as "thoughts," Proclus claims that this discussion concerns the question of where the Forms are situated, in mind or prior to the mind? The subsequent discussion shows, in Proclus's view, the difficulties involved in either alternative. After a lengthy and obscure commentary of the Platonic passage, Proclus concludes that Plato intends to demonstrate that participation must be understood neither as a corporeal process, nor as a physical one, nor as an intellectual (psychic) one.[Procl. in Parm. 905 - 906; Morrow, 265]

In the examination of the last part of Plato's passage, where Forms are interpreted as paradigms standing in Nature, Proclus highlights the distinction between the unchangeable Forms that stand "fixed" and the sensible things that come to be depending on them.[Procl. in Parm. 908; Morrow, 266]

Proclus focuses on the new mode of participation, called by Plato "assimilation" (ὁμοίωσις). In Proclus's view, it has a double function: on the one hand, it is intended to avoid the problem whether the particular things participate in the Forms either as wholes or as parts [Procl. in Parm. 906, 912; Morrow, 265, 269], and, on the other hand, to remove the difficulty arising from the assumption that Forms are coordinate with the sensible things.

> "For the paradigm is not present to the image, nor is it coordinate with it. Participation, then, takes place by assimilation. He has introduced this concept by calling the Forms patterns and the things participating them likenesses, and this participation for this reason assimilation."[Procl. in Parm. 906; Morrow, 265]

According to the new semantics, the Forms are paradigms (παραδείγματα) and the sensible things are "copies," "images" (εἰκόναι) or "likenesses" (ὁμοιώματα) of them. Thus, participation is defined to be equivalent to "assimilation".[Procl. in Parm. 906 - 907; Morrow, 265 - 266] Proclus considers that Plato is confident that this new definition of participation is correct; he states that

> " 'it appears to him very likely' that this is the method of participation".[Procl. in Parm. 907; Morrow, 265]

On the other hand, Proclus is aware that the mere introduction of a correct definition of participation does not solve the Third Large paradox:

> "Socrates' present hypothesis constitutes an advance, without constituting a complete solution."[Procl. *in Parm.* 907; Morrow, 266]

The solution is viewed by Proclus in Plato's definition of the concept of "similarity". Proclus explicitly declares that the similarity relation between a thing and the Form is not reciprocal

> "even if the sensible object is "similar to"[7] the intelligible Form (εἰ τὸ αἰσθητὸν ὅμοιόν ἐστι πρὸς τὸ νοητὸν εἶδος), it is still not necessary that this relationship be reciprocal, and the Form be said to be "similar to" it" (οὐ χρὴ καὶ ἀντιστρέφειν τὴν σχέσιν κἀκείνῳ τοῦτο λέγειν ὅμοιον).[Procl. *in Parm.* 912; Morrow, 270]

Proclus explains that if the similarity is taken to be reciprocal, then a new infinite regress will appear, the so-called "resemblance regress"

> "lest prior to these two "like" entities we should go in search of some further Form as cause of the similarity[8] between these two (ἵνα μὴ πρὸ τῶν δύο ὁμοίων ἀλλήλοις ἕτερόν τι πάλιν ζητήσωμεν εἶδος, ἀμφοῖν τῆς ὁμοιότητος αἴτιον)· for things that are like one another have in every case in common some identical quality, and it is in virtue of this identical quality in them that they are said to be "like" (τὰ γὰρ ἀλλήλοις ὅμοια πάντως ταὐτοῦ τινος κεκοινώνηκε, καὶ δι' ἐκεῖνο ταὐτὸν ἐν αὐτοῖς ὂν ὅμοια λέγεται εἶναι). Once this has been agreed, then, that the participated entity and the participating, the paradigm and the image, are like one another, there will be prior to these another thing that causes their similarity, and this progression will continue to infinity" (Τούτου γὰρ συγχωρηθέντος, ἀλλήλοις ὅμοια εἶναι τό τε μεθεκτὸν καὶ τὸ

---
[7]Morrow uses the term "like" instead of the term "similar."
[8]Morrow uses the term "likeness" instead of the term "similarity".

μετέχον, τὸ παράδειγμα καὶ τὸ εἰκασθὲν, ἔσται καὶ πρὸ τούτων ἄλλο τι τὸ ταῦτα ἀφομοιοῦν, καὶ τοῦτο ἐπ' ἄπειρον).[Procl. in Parm. 912; Morrow, 270]

In this connection, Proclus makes an important distinction. He clarifies that the term "similar" (τὸ ὅμοιον) is used in two senses (τὸ ὅμοιον εἶναι διττόν):

1) "similarity" denotes a relation between two *coordinate* entities (τὸ μὲν σύζυγον ὁμοίῳ ὅμοιον);

2) "similarity" denotes the *subordination* of a thing to an archetype (τὸ δὲ ὡς ὑφειμένον πρὸς ἀρχέτυπον ὅμοιον).

Proclus further clarifies this distinction stating that similarity in the first sense consists in the identity of a reason-principle among coordinate thing, whereas in the second sense, similarity involves not only identity but also

"otherness, whenever something is "similar" as having the same Form derived from the other, but not *along with it*. [Procl. in Parm. 912; Morrow, 270]

i.e., the relation in the second sense holds between a thing and a Form, that is between entities of different levels. Proclus emphasises that

"it is false that reciprocity extends to all levels alike" (εἶναι ψεῦδος τὸ ἐπὶ πάντα ἁπλῶς τὴν ἀντιστροφὴν ἐκτείνειν).[Procl. in Parm. 914; Morrow, 272]

Further, Proclus examines the question posed in Plato's *Parmenides*, if the sensible things are "similar" to the Forms, will the Forms necessarily be "similar" to them? Proclus states from the very outset that an affirmative answer to this question is false [Procl. in Parm. 914; Morrow, 272] and points to the significance of the assumption

"like things participate in one and the same Form, in virtue of which their similarity[9] exists" (τὰ ὅμοια ἑνὸς ταὐτοῦ τε

---

[9]Morrow uses the term "likeness" instead of the term "similarity".

μετείληφεν εἴδους, καθ' ὃ ἡ ὁμοιότης).[Procl. in Parm. 915; Morrow, 272]

Proclus concludes that the Platonic thesis "neither the Form is similar to a participant of it, nor the participant similar to the Form"

"brings the argument round to a refutation" (οὕτως ὁ Παρμενίδης εἰς ἔλεγχον περιάγει τὸν λόγον).[Procl. in Parm. 918; Morrow, 274]

and thereby concludes that it is not in virtue of similarity that the sensible things participate in the Forms, but

"in virtue of another, more powerful cause – a more correct conclusion" (καὶ συνάγει λοιπὸν ὡς οὐ διὰ τὸ ὅμοιον τὰ τῇδε μετέχει πάντων τῶν εἰδῶν, ἀλλὰ δι' ἄλλην κυριωτέραν αἰτίαν, ἀληθέστερον εἰπών).[Procl. in Parm. 918; Morrow, 274]

In conclusion, Ploclus's treatment of the Third Large argument shows that he considers it as

a) the same as the Third Man argument.

b) resolved by Plato in his *Parmenides*.

However, in his analysis Proclus does not use the language specific to the Peripatetic philosophers, notably, he does not use the concept of predication, but focuses on the analysis of Plato's discourse and the explication of his metaphors. In particular, he makes important clarifications concerning the double sense of the concept of similarity and the reciprocity of the relationship between a sensible thing and a Form in which it participates. He makes clear that similarity is a relation between coordinate entities, whereas participation is a relation between entities of different levels. If the concept of similarity is applied to entities of different levels, then the scheme of ascension from sensible particulars to their Forms becomes vicious.

6 On the correctness of the argument and its solution.

How to check the correctness of the Third Large / Man argument? How to check the consistency of Plato's theory of Forms or Aristotle's theory of categories? Pelletier and Zalta have suggested an answer to this question.

> "one way to show that a theory is consistent and worthy of being taken seriously is to try to reconstruct it using modern logical methods". [24, p. 198]

Accordingly, they analyse Plato's Third Large argument in a comprehensive axiomatic theory of abstract objects advanced in [33], which axiomatizes two primitive modes of predication:

(i) the formula 'Fx' is interpreted as x *exemplifies* F or x instantiates F, and

(ii) the formula 'xF' is interpreted as x *encodes* F or F is constitutive of x (or F is part of the nature of x).

This distinction is inspired by Meinwald's studies on Plato's *Parmenides* [21, 20], where two types of predication are distinguished:

a) predications of the kind 'x is F *pros ta alla*', and

b) predications of the kind 'x is F *pros heauto*.'

The first type is the ordinary predication when x is an object and F is a property which x exemplifies or instantiates in the traditional sense, whereas the latter type is a special mode of predication which holds when x is a Form and F is a property which is, in some sense, part of the nature, definition or concept of the Form.

In this theory, sensible objects can only exemplify their properties, whereas ideal entities both exemplify properties and encode them. The properties that ideal entities encode, unlike the properties that they exemplify, are the ones by which they are individuated. The theory is supplied with a comprehension principle which states under what conditions there are ideal entities that encode properties and an identity principle, that provides a criterion of identity of ideal objects.

This theory provides the required general framework to develop a Platonic theory of Forms and find a complete solution to the Third Man argument. This general theory is "typed" and thereby is proved consistent. Forms are distinguished from properties because the latter are predicable entities, whereas the Forms are not. In particular, this theory avoids Russell's paradox. The authors note that Plato's interpreters who identify Forms and properties, such as Vlastos and Meinwald,

> "on the grounds that this is the most sympathetic understanding of Plato, owe us a consistency proof of their respective theories of Forms. Without such a proof, *our* theory may be the more sympathetic understanding of Plato." [24, p.182]

In our work [30], we adopted a different approach. We represented Plato's line of reasoning in a formal language of Frege-Russell type. The language was constructed in such a way that the behaviour of Forms as ordinary particulars to be expressible in this language. This required the introduction of the *confusion condition*

> If x is an individual variable and φ a formula, then $\{x|\varphi\}$ is a *term* (the *confusion condition*).

that allows treating predicates as bindable, like individuals [30, p.42]. This language cannot be a first-order language, because of the 'confusion' of particulars with the *eide*, which gives rise to the infinite regress in both Platonic and Peripatetic versions. The language is supplied with a *reducibility scheme*, which expresses the equivalence of participation with the attribution of a predicate and a *principle of extensionality*, that guarantees the individuation of the *eide*. In symbolic language these principles are stated as follows:

(P1)   *Reducibility scheme:* $x \in \{z|F(z)\} \equiv F(z|x)$

(P2)   *Principle of extensionality:* $((\forall x)(F(x) \equiv G(x))) \supset F = G$

(D1)   *Individuation of the Forms:* $F = G \underset{Def}{\longleftrightarrow} (\forall x)(F(x) \equiv G(x))$

Following Plato's difference in the use of the terms *eidos* and *idea*, we distinguish Forms from properties. In this language, Plato's argument can be formally represented step by step, giving rise to an infinite sequence of Forms

$$\begin{cases} \{z|F_n(z)\} \cup F_n, \ n = 0, 1, 2, 3, \ldots \\ F_0(x) \equiv x \in \{z|F_0(z)\} \end{cases}$$

that have the property

$$F_n(x) \equiv F_0(x) \vee (x = \{z|F_0(z)\}) \vee \ldots \vee (x = \{z|F_{n-1}(z)\})$$

In this way, one and the same *idea* (being large or being a man) corresponds to an ascending sequence of *eide* with expanding extents, which contradicts the individuation of Forms (D1). This language is not consistent; in particular, Russell's paradox is derivable in it.

The definition of *eide* as paradigms found in nature compels us to modify the initial language and reformulate the reducibility scheme by introducing a second-order form of comprehension scheme[10].

(P1*) *Principle of comprehension*: $(\exists F)(\forall x)(x \in F \equiv F(x))$

The new scheme enables us to define instantiation as equivalent to participation (satisfaction of predicate). The two-place predicate of similarity that compares Forms with its instances, is introduced by the notation

$$H(\_, \ldots) \leftrightarrow \_ \text{ is similar to } \ldots$$

In particular, the predicate $H(\{z|F(z)\}, x)$ is read: the Form $\{z|F(z)\}$ is similar to the (fixed) individual x that instantiates F. Insofar x is a fixed individual, the predicate H depends on the Form $\{z|F(z)\}$. Specifically, on the predicate F. Then, the following question is posed: Is it true that

$$(\forall F)[F(x) \supset H(\{z|F(z)\}, x)?$$

---

[10]The reducibility scheme (P1) that we use in the formulation of the Third Large Paradox lacks an existence claim. Plato explicitly introduces the existence claim for the Forms in the passage where he states that the Forms exist in nature. However, the Third Man Paradox can be comfortably formulated without using an existential claim. For this reason, we adopted a reducibility scheme for the formulation of the argument but introduced the second-order scheme (P1*) when Plato makes an explicit existence claim for the Forms.

A positive answer gives rises to the so-called *resemblance regress*, because if the proposition $H(\{z|F(z)\},x)$ holds for all predicates F, then the Form $\{z|F(z)\}$ itself is allowed to occur among the values of F. Hence, the class of all possible predicates F must be restricted. This is attained by the definition of similarity, according to which if $H(\{z|F(z)\},x)$ holds, then the object z must participate in the same Form in which the instance x participates, that is $F(z|x)$, which denotes the result of the substitution of the fixed object x everywhere in z. This suggests the following sound definition of similarity

$$H(\{z|F(z)\}, x) \underset{Def}{\longleftrightarrow} F(z|x).$$

The domain of the predicate H is the collection of all arguments *homogeneous* (similar to) the fixed instance x, within instantiation. In other words, the domain of H includes all those values z that satisfy the same predicate F, that x instantiates. Thus, H defines a domain of objects homogeneous to x. In this case, the Form $\{z|F(z)\}$ is not allowed to occur in the domain of H.

Hence, Forms and particulars are not homogeneous entities and the relation of participation cannot be defined by the predicate of similarity, as stated in Plato's text.

Consequently, our modified with the principle (P1*) language, supplied with the sound definition of similarity, according to Plato's texts, is typed and consistent. This is a complete solution to the Third Large Paradox.

We have thus demonstrated that there exists at least one model in which Plato's reasoning is representable and true (Zalta's language is another model). This refutes Scaltsas's strong claim that there is a necessary falsehood in the Third Man Argument in [29]. If this claim is understood as stating that the Third Man Argument is false in all models, then it is false.

7 Type restrictions and Greek mathematics. Plato and Eudoxus.

Larry Hickman showed in [14] that there was a late scholastic theory of types connected with problems of simply non-descending predicates.

These problems were faced by the elaboration of an informal technique of elimination of paradoxes involving self-reference (the so-called *paradoxes of descent*) based on the distinction in types of predication corresponding to two senses of the verb 'to be', one *in sensu compositi* and another *in sense divisi*. The examination of the scholastic variety of theory of types led him to the conclusion that the

> "simple theory of types was in the offing from the time of Plato and Aristotle".[14, p. 180]

By 'simple theory of types' we have to understand the fundamental idea that not all entities are of the same kind, in terms of logical function, i.e., that some properties can be meaningfully predicated of certain entities but not of others. As we have shown in the previous sections of this paper, Plato and Aristotle seem to be aware of this principle; moreover, they were aware that its violation leads to paradoxes of the type of 'Third Man'.

In this section, we turn our attention to Greek mathematics to examine if the Greek mathematicians follow in their reasoning the restrictions imposed by the above principle of the simple theory of types. This important question was posed for the first time and examined systematically by Jean-Louis Gardies in [12].[11] Gardies clarifies that he adopts a deliberately anachronistic approach to check if the Greek mathematicians keep the restrictions of the theory of types, which was shaped at the end of the 19th century by Peano, Frege and assumed a definite form at the beginning of the 20th century by Russell. Thus, he uses the term "simple theory of types" in the sense of *Principia Mathematica* [32, p. 39 - 68] and its subsequent interpretations by David Hilbert [9, 10]. Concerning the Greek philosophers, notably Plato and Aristotle, this anachronistic approach was first adopted by Bertrand Russell who has argued that Plato violates in his arguments the restrictions imposed on language by the theory of logical types [28, pp. 126 - 130]. Nevertheless, the question makes sense in the wider context of Universal Logic understood as a meta-logical approach that aims at addressing and formalizing the nature of what may be called 'logic' as a form of

---

[11]See also Gardies' review of [14] in [13].

"sound reasoning" and identifying possible principles underlying sound reasoning in different domains, developed under different logics.[23]

Gardies argues that the mathematical discourse of the works of notable Greek mathematicians (Euclid, Archimedes, Apollonius, and others) is essentially first-order, i.e., deals with objects of our immediate, most elementary experience of sensitive objects that the Aristotelian tradition would qualify as *primary substances* and today as predicates of n arguments, which all refer to individuals. However, Euclid in his *Elements* does not confine himself to the use of predicates of Aristotelian form only, i.e., one-place predicates (e.g., when it is stated that an object is a point, or line, etc.). He also uses two-place predicates, when it is stated that an object is smaller or greater than or equal to another object or a line is parallel or perpendicular to another line. He uses three-place predicates when refers to the relation ... *is between* ... *and* ... applied, for instance, to three distinct points of the same straight line. Thus, in the first four Books, reasoning does not go beyond the first-order logic that involves predicates of individuals of n arguments.

Only in Book V, where Euclid exposes Eudoxus's general theory of proportions, he encounters an object of different nature than the individuals that occur in the first four Books. This is the concept of *ratio* (λόγος) between *two homogeneous* (ὁμογενῶν) *magnitudes with respect to their size* (Definition 3). It is a specific relation between magnitudes of the same kind. A ratio is a second-order object of one first-order object to another first-order object. How Euclid defines this object? Definition 5 of Book V states that

> Magnitudes are said to *be in the same ratio*, the first to the second and the third to the fourth, when, if any equimultiples whatever are taken of the first and third, and any equimultiples whatever of the second and fourth, the former equimultiples alike exceed, are alike equal to, or alike fall short of, the latter equimultiples respectively taken in corresponding order. [8, pt. II, p. 114]

The sameness of ratios (the *definiendum*) expressed at the beginning of the sentence concerns objects which are not first-order, whereas the definiens expressed in the second part of the sentence ("when, if any equimultiples whatever ...") concerns first-order objects. In this way, Euclid succeeds to define the equality of second-order objects (the ratios) by using a four-place predicate of individuals, i.e., of first-order objects.

In an analogous, although not identical way, the two-place predicate H of similarity which Plato introduces to solve the "Third Large" paradox, is not a first-order object, since it depends on the *eidos F*, is defined by restricting F only to those individuals that instantiate F, i.e., confining himself to first-order objects.

What is interesting in the case of Plato and Eudoxus is that they were able to go beyond the level of reasoning concerning first-order objects that they handled by resorting to the sameness of ratio and the instantiation of "one and the same" predicate, respectively. The most remarkable thing is that the definitions suggested by Plato and Eudoxus use *definienda* that remain within the limits of the first-order logic. Both Plato and Eudoxus imagine objects which are not first-order but succeed to define them remaining within the limits of first-order logic.

Consequently, it seems that Plato, in dealing with the "Third Large" paradox, follows a practice of reasoning established by mathematicians of the 5th century, notably by the genius of Eudoxus.

8 Conclusion

We examined the problem of self-reference paradoxes of the type of "Third Man" that appear first in Plato's *Parmenides*, and further discussed in Aristotle, and the Peripatetic commentators, and Proclus.

We showed that the various versions were analysed using different language peculiar to the Peripatetic and Neo-Platonic schools. Their analyses and perspectives reflect different understandings by Plato, Aristotle, and their corresponding commentators. We showed that the Peripatetic commentators do not focus on Plato's solution but primarily on the statement of the paradox and attempt to reveal its implicit assumptions. On the contrary, Proclus seems to be convinced that Plato suggests a sound solution to the paradox by the definition of the predicate of similarity (homogeneity) that demarcates two types of homogeneous entities – the eide and the participants in them, in a way that their confusion would be inadmissible.

We also claim that the Peripatetics had no reason to examine the Third Man paradox in terms of Aristotle's theory of categories if the paradox was not relevant to the latter. If the paradox was perceived as concerning only Plato's theory of Forms, they would have had presented it as a fatal argument against its consistency, i.e., as a convenient argument against a rival theory. In this case, they would have had no reason to introduce restrictions upon predication in the theory of categories.

We demonstrate that Plato's solution follows a sound line of reasoning that is formalizable in a language of Frege-Russell type; hence there exists a model in which Plato's reasoning is valid.

Furthermore, we notice that Plato's definition of the second-order predicate of similarity is attained by resorting to first-order objects, i.e., particular things. In this sense, Plato's definition is comparable to Eudoxus' definition of ratio in Book V of Euclid's *Elements*, which is also attained by resorting to first-order objects. Consequently, Plato seems to follow a logical practice established by the mathematicians of the 5th century, notably Eudoxus, in his solution to the "Third Large" paradox.

References

[1] Alexander of Aphrodisias 1891. *Commentaria in Aristotelem Graeca. Commentary to Aristotle's Metaphysics* Michael Haydock (Ed.), Berlin.

[2] Allen, R.E. 1983. *Plato's Parmenides: Translation and Analysis.* University of Minnesota Press.

[3] Aristotle, Becker, I (Ed.) *Aristotelis Opera Omnia.* Berlin, 1831-1870. New ed., Gijon O. (Ed.), 1960-.

[4] Barnes, Jonathan (General Editor). 2014. *The Complete Works of Aristotle*, 2 vols, 1984; reprinted with corrections, 1995.

[5] Brommer, Peter, 1940. *Eidos et Idea. Étude Sémantique et Chronologique Des Oeuvres de Platon.* Assen : Van Gorcum.

[6] Dorter, Kenneth, 1994. *Form and Good in Plato's Eleatic Dialogues. The Parmenides, Theaetetus, Sophist, and Statesman.* Berkeley, Los Angeles, Oxford: University of California Press. Available online: http://ark.cdlib.org/ark:/13030/ft2199n7gn/

[7] Fine, Gail, 1993. *On Ideas. Aristotle's Criticism of Plato's Theory of Forms.* Oxford: Clarendon Press.

[8] Heath, Thomas L. 1956. *The Thirteen Books of Euclid's Elements*. Dover Publications.

[9] Hilbert, David and Bernays, Paul, 1934-39. *Grundlagen der Mathematik*, Berlin: Springer, vol. 1, 1934, vol. 2, 1939; 2e ed, 1968-1970. trad. fr. by F. Gaillard, E. and M. Guillaume, Paris: L'Harmattan, 2001.

[10] Hilbert, David and Ackermann, W. 1928. *Grundzüge der theoretischen Logik*, Berlin: Springer. 5th ed., 1967.

[11] Gazziero, Leone. 2014. «ἐὰν ὡσαύτως τῇ ψυχῇ ἐπὶ πάντα ἴδῃς (Platonis Parmenides, 132a 1 - 132b 2) Voir Les Idées Avec Son Âme Et Le « Troisième Homme » De Platon», *Revue De Philosophie Ancienne*, XXXII (1), 35-85.

[12] Gardies, Jean-Louis. 2005. « Les mathématiques grecques sous le regard de la théorie des types », *Philosophia Scientiæ*, cahier special 5, 11-25.

[13] Gardies Jean-Louis. 1982 Review of [14], *Revue Philosophique de la France Et de l'Etranger* T. 172, No. 1 (Janvier-Mars1982), 104-105.

[14] Hickman, Larry, 1980. *Modern Theories of Higher Level Predicates. Second Intentions in the Neuzeit*. Munchen. Philos. Verlag.

[15] Kraut, R. (ed.). 1992. *The Cambridge Companion to Plato*, Cambridge: Cambridge University Press.

[16] Liddell H.G., Scott R. (ed. Jones H.S.), 1925-40. *Greek-English Lexicon*. Oxford.

[17] Losev, Aleksey F. 1993. *Essays in Ancient Symbolism and Mythology*. Moscow, first edition 1930; Reprint: Moscow: Mysl', 1993 [in Russian].

[18] Losev, Aleksey F. 1963-88. *History of Ancient Aesthetics*. 7 vols. Moscow: Isskustvo [in Russian].

[19] Lloyd P. Gerson, 2011. "Proclus and the Third Man", *Études platoniciennes*, 8, 105-118.

[20] Meinwald, C. C. 1992. 'Good-bye to the Third Man', in [15]: 365–396.

[21] Meinwald, C. C. 1991. *Plato's Parmenides*, New York: Oxford University Press.

[22] Mignucci, Mario. 1990. "Plato's 'Third Man Argument' in the Parmenides." *Archiv für Geschichte der Philosophie* 72: 143–81.

[23] Mossakowski, Till; Joseph Goguen, Răzvan Diaconescu and Andrzej Tarlecki 2007. "What is Logic?" J.-Y. Beziau (Ed.), *Logica Universalis*, 2nd edition, 111–133 2007 Birkhäuser Verlag Basel, Switzerland.

[24] Pelletier, Francis Jeffry and Edward N. Zalta. 2000. "How to Say Goodbye to the Third Man," *Noûs* **34**(2), 165–202.

[25] Rickless, Samuel, "Plato's *Parmenides*", *The Stanford Encyclopedia of Philosophy* (Spring 2020 Edition), Edward N. Zalta (ed.),

URL = <https://plato.stanford.edu/archives/spr2020/entries/plato-parmenides/>.

[26] Rimondi, Giorgia, 2018. "Losev's analysis of Plato's terminology: on the notions of *eidos* and *idea*," *Platonic Investigations*, VIII (1), 73-85. Moscow, Saint-Petersburg: Plato Philosophical Society.

[27] Ross, W.D. (General Editor). 1950. *The Works of Aristotle*. Vol. I. London: Oxford University Press. 1928; repr. 1937, 1950.

[28] Russell, Bertrand 1972. The History of Western Philosophy. New York: Simon Schuster.

[29] Scaltsas, T. 1992. "A Necessary Falsehood in the Third Man Argument." *Phronesis*, 37: 216–232.

[30] Vandoulakis I.M. 2009. "Plato's Third Man Paradox: its Logic and History," *Archives Internationale d'histoire des Sciences*, Vol. 59 No 162, 3-53.

[31] Vlastos, Gregory, 1954. "The Third Man Argument in the Parmenides." Philosophical Review 63, 319-349.

[32] Whitehead, A.N. and Russell, B. 1910. *Principia Mathematica*, Cambridge, vol. 1, 1910, vol. 2, 1912, vol. 3, 1913; 2nd ed., 1925-1927.

[33] Zalta, Edward N. 1983. *Abstract Objects: An Introduction to Axiomatic Metaphysics*. Synthese Library, 160. Dordrecht, Netherlands: D. Reidel Publishing Company.

Stoic Natural Deduction and Intuitionistic Stoic Logic

Victor Nascimento
Rio de Janeiro State University
`victorluisbn@gmail.com`

Abstract

This paper has a twofold aim. First, we show that Bobzien's sequent calculus for Stoic logic can be transformed into a system of natural deduction on which four of the three *themata* are absorbed into the structure of derivations. Since no benefits seem to arise from the restriction imposed by the Stoics on compositionality, an unrestricted version of Stoic logic is also proposed. Second, we show that Stoic logic can easily be modified to generate new systems with remarkable properties. To exemplify, we define a system called intuitionistic Stoic logic, a proper subsystem of intuitionistic logic which has many interesting properties that it lacks — such as being relevant and paraconsistent but, unlike minimal logic, allowing use of the disjunctive syllogism.

1 Introduction

A great injustice has long been perpetrated by historians of logic against Stoic philosophers, either in the form of outright dismissal of their contributions to the development of logic or of erroneous attribution of such advancements to Aristotle. As pointed out by eminent polish logician Jan Łukasiewicz, historical textbooks did not mince words in their criticism of Stoic logic:

I am thankful for Prof. Luiz Carlos Pereira's suggestions, revision and guidance. This study was financed in part by the Coordenação de Aperfeiçoamento de Pessoal de Nível Superior – Brasil (CAPES) – Finance Code 001.

> Maier says that the Stoic Logic, which in fact is a masterpiece equal to the logic of Aristotle, yields a poor and barren picture of formalistic–grammatical unsteadiness and lack of principle, and adds in a footnote that the unfavourable judgement of Prantl and Zeller on this logic must be maintained. The *Encyclopaedia Britannica* of 1911 says briefly of the logic of the Stoics that "their corrections and fancied improvements of the Aristotelian logic are mostly useless and pedantic". [15, p.49]

The complete inaccuracy of those accounts is now evident not only due to the extensive works of Łukasiewicz, but also to the careful reexamination of original texts by contemporary historians. In [16], for example, Benson Mates shows that many fundamental technical aspects of modern logic were already being discussed by the Stoics and their predecessors, the dialeticians of the Megarian school. Aside from anticipating truth tables, the dialetician Philo of Megara had already given a definition of what is nowadays known as the material implication, more than two millennia before its rediscovery by Frege [8]; Diodorus Cronus objected to the use of material (Philonian) implication and argued for its substitution by a modalized conditional operator, which is akin to many modal implications created only after the formalization of modal logic by Barcan Marcus [1], Lewis [14] and Kripke [13] [16, pg.42–51]. Both accounts were further discussed and built upon by Stoic logicians such as Chryssipus, and Stoic logic itself was developed by considering strengths and shortcomings of those conceptions.

It is not without reason, then, that Mates joins Łukasiewicz in his harsh criticism:

> The period of Aristotelian dominance in logic might well have ended sooner if certain ancient texts had been studied more carefully.(...) Comparing such facts with the extremely adverse and inaccurate characterizations of Stoic logic by Prantl, Zeller, and other "standard" authors, and observing that a similar situation obtained with respect to medieval logic, Łukasiewicz understandably came to the conclusion that the history of logic ought to be rewritten. [16, pp. 2–3]

Many questions which are still open today were already being vividly and fruitfully discussed by the Stoics, and the reexamination of those debates started recently enough for us to conclude that there is much yet to explore. Moreover, recent work on the subject have successfully laid out the ground for such an endeavor. With regards specifically to proof theory, a thorough reconstruction of Stoic logic[1] was provided by Bobzien in [3], which is formulated using the structure of Gentzen's sequent calculus [9]. The purpose of this paper is to provide some modest contributions to the advancement of Stoic proof theory, and also to show how such an advancement can possibly benefit seemingly unrelated areas of logic.

Section 2.1 provides a brief introduction to Bobzien's Stoic sequent calculus. In section 2.2 it is shown that three of the four Stoic *thematas* are naturally absorbed by structural features of natural deduction. In section 2.3 we present a system of Stoic natural deduction which is faithful to Bobzien's Stoic sequent calculus, but whose structure slightly deviates from standard systems of natural deduction. Section 2.3 contains a brief discussion of Stoic logic's restriction on compositionality, which is followed by the definition of a variant of Stoic natural deduction which does not strictly corresponds to Bobzien's sequent calculus, but which satisfies unrestricted compositionality and is much more similar to standard systems of natural deduction. Section 2.5 briefly discusses some problematic aspects of constructive implication before defining a constructive version of Stoic natural deduction and showing that it satisfactorily avoids such problems. Finally, Section 3 provides some closing remarks about important structural features observed in Stoic systems.

---

[1]It is important to notice that "Stoic logic" is not a univocal term, and its use can be made in reference both to the Stoic's original system and to Bobzien's (or other author's) reconstructions. This paper deals mainly with the latter, and any further reference to "Stoic logic" should be read specifically as a reference to Bobzien's reconstruction.

2 Stoic Sequent Calculus and Stoic Natural Deduction

2.1 Bobzien's Sequent Calculus

Some conventions adopted by Bobzien will be slightly altered. We use arrows ($\Rightarrow$) instead of turnstiles to separate the antecedent and consequent of sequents, and $\vdash$ will be reserved for expressing derivability relations of a given system.

The definition of rules must be preceded by a formalization of the Stoic notion of contradictories (*antikeimena*), in which two propositions are said to be contradictories if one "exceeds the other by a negation":

Definition 2.1. *We define the contradictories A^* of a proposition A as follows*[2]:

1. For propositions A whose main operator[3] is not negation, A^* is $\neg A$;

2. For propositions $\neg A$, $(\neg A)^*$ is either A or $\neg\neg A$.

We then define rules for the system, stipulating that Γ and Δ are sets[4] and that the operators $\wedge$, $\rightarrow$ and $\oplus$ represent, respectively, *conjunction*, *implication* and *exclusive disjunction*:

Themata

$$\frac{\Gamma, A \Rightarrow B}{\Gamma, B^* \Rightarrow A^*} \text{ T1} \qquad \frac{A, B \Rightarrow C \quad C, \Delta \Rightarrow D}{A, B \Rightarrow D} \text{ T2}$$

---

[2] A^* is being taken as a variable; the symbol * must not be read as a proper operator. Moreover, parentheses are used on the second clause solely to avoid ambiguity, as $\neg A^*$ could be read both as $(\neg A)^*$ and $\neg(A^*)$. We would like to thank one of the reviewers for pointing this out.

[3] For a formal definition of main operators, see the definition of "main functor" in [5, pg. 22].

[4] Bobzien uses the notion of "multisets without repetition" in order guarantee that there are at least two distinct formulae on the antecedent of any sequent, but we can use sets whenever this feature is guaranteed by other means.

Stoic Natural Deduction

$$\frac{A, B \Rightarrow C \quad C, \Gamma \Rightarrow D}{A, B, \Gamma \Rightarrow D} \text{ T3} \qquad \frac{A, B \Rightarrow C \quad C, \Delta, \Gamma \Rightarrow D}{A, B, \Gamma \Rightarrow D} \text{ T4}$$

Restriction on T1: $B^* \notin \Gamma$.
Restriction on T2 and T4: $\Delta \subseteq \{A, B\}$.
Restriction on T3: $A \notin \Gamma$ and $B \notin \Gamma$.
Restrictions on all Themata: Γ and Δ cannot be empty, and there must always be at least two distinct formulae to the left of $\Rightarrow$.

Primitive Indemonstrables

$$\frac{}{A \to B, A \Rightarrow B} \text{ A1} \qquad \frac{}{A \to B, B^* \Rightarrow A^*} \text{ A2}$$

$$\frac{}{\neg(A_1 \wedge A_2), A_i \Rightarrow A_j^*} \text{ A3} \qquad \frac{}{A_1 \oplus A_2, A_i \Rightarrow A_j^*} \text{ A4}$$

$$\frac{}{A_1 \oplus A_2, A_i^* \Rightarrow A_j} \text{ A5}$$

Derived Indemonstrables

$$\frac{}{A, B^* \Rightarrow \neg(A \to B)} \text{ A6} \qquad \frac{}{A, B \Rightarrow A \wedge B} \text{ A7}$$

$$\frac{}{A, B \Rightarrow \neg(A \oplus B)} \text{ A8} \qquad \frac{}{A^*, B^* \Rightarrow \neg(A \oplus B)} \text{ A9}$$

Restriction on A3, A4 and A5: $i, j \in \{1, 2\}, i \neq j$.
Restriction on Themata T1–T4 and Indemonstrables A1–A9: the same formula cannot occur twice on the antecedent of the same sequent.

In Stoic sequent calculus, the role of *axiom rules* is fulfilled by the *indemonstrables*, and the role of *inference rules* is fulfilled by the *themata*. Indemonstrables A6 through A9 can be used freely but need not be explicitly added to the system[5], as they can be derived/analyzed using

---

[5] A2 can be analyzed into A1 and T1 and is also redundant; we define it explicitly for purely conventional reasons.

the primitive indemonstrables and the *themata*. Although this system is useful for proofsearch and the obtainment of technical results, it is not ideal if one wants to manually explore valid derivations. As is often the case with formulations using sequent calculus, the structure is perfectly functional but relatively unintutive.

Since natural deduction is usually better suited to represent intuitions underlying a logical system, a reformulation of Stoic logic in it could be pedagogically more sound than one in sequent calculus. In fact, as one would expect, the correspondence between cut rules and composition makes it so that only indemonstrables and the first *thema* are needed for derivations. The three remaining *themata* lose their status as proper rules and may now be seen as mere descriptions of the derivation's structure, which are already incorporated in the definitions of what counts as a deduction. Before defining the new natural deduction system, we will briefly show the specific manner in which each *thema* is absorbed by the structure of derivations, as well as discuss why this does not work for the first *thema*.

2.2 Absorption of the *Themata*

The first *thema* is informally stated as follows[6]:

First Thema. *When from two propositions a third is deduced, then from either of them together with the/a contradictory of the conclusion the/a contradictory of the other is deduced.*

Extended First Thema. *When from two or more propositions another is deduced, then from all but one of them together with the/a contradictory of the conclusion the/a contradictory of the remaining one is deduced.*

As noted by Bobzien, the first *thema* is closely related to the $L\neg$ and $R\neg$ rules of Gentzen's sequent calculus, aside from being the main rule for dealing with negation. Furthermore, when combined with the notion

---

[6]Since the extended version of each *thema* is always more general than the initial version, only the former is used for defining rules. The original version of each *thema* defines a *schema* for all applications of the extended *thema* which use the least possible amount of premises, and are thus always contained in the definition of the extended version.

of contradictories, it is also the main rule for applying the inference of *reductio ad absurdum*. In our reformulations, it is the only *thema* which will not be absorbed by the calculus, as well as the only rule which discharge hypotheses:

$$\begin{array}{cc} \Gamma & [A] \\ \vdots & \\ C & C^* \\ \hline & A^* \end{array} T, A$$

Restrictions: there must be at leat one assumption with shape A above C, discharge of all assumptions with shape A is mandatory and Γ must not be empty.

Γ is not empty, C is being deduced from A and at least one formula from Γ. To this derivation we may add the assumption C^* and apply the rule to conclude A^* from only Γ and C^*, since A is discharged by the rule.

Extended Third Thema. *When from two propositions a third is deduced, and from the third and one or more external propositions another is deduced, then this other is deduced from the first two and the externals.*

This *thema* can be interpreted as a statement of the property of *restricted compositionality*:

$$\begin{array}{c} A \quad B \\ \Pi^1 \\ C \end{array} + \begin{array}{c} C \quad \Gamma \\ \Pi^2 \\ D \end{array} \Rightarrow \begin{array}{cc} A \quad B & \\ \Pi^1 & \\ C & \Gamma \\ \Pi^2 & \\ D & \end{array}$$

Restriction: Π^1 only has A and B as open assumptions, and Γ contains neither A nor B.

After composing the two derivations, we obtain a deduction of D which depends only on A, B and Γ, as desired. However, since the first derivation must always have exactly two open assumptions on its dependencies, compositionality is *restricted*.

Second Thema (reconstructed): *When from two propositions a third is deduced, and from the third and one – or both – of the two another is deduced, then this other is deduced from the first two.*

Fourth Thema (reconstructed): *When from two propositions a third is deduced, and from the third and one – or both – of the two and one or more external propositions another is deduced, then this other is deduced from the first two and the externals.*

The *themata* are absorbed by the same aspect, which is consistent with Bobzien's expositions of criticism formulated by peripatetics [3]. They can be interpreted as defining that Stoic logic uses *sets* of formulas in its definitions:

$$
\begin{array}{c}
A_1 \quad A_2 \\
\Pi^1 \\
C
\end{array}
\quad + \quad
\begin{array}{c}
C \quad A_i \\
\Pi^2 \\
D
\end{array}
\quad \Rightarrow \quad
\begin{array}{cc}
A_1 \quad A_2 & \\
\Pi^1 & \\
C & A_i \\
\Pi^2 & \\
D &
\end{array}
$$

Restriction: $i \in \{1, 2\}$, and Π^1 only has A_1 and A_2 as open assumptions.

Even though there is now a new occurrence of either the assumption A^1 or the assumption A^2 on the derivation, the *themata* tells us that it should not be counted, validating the notion that multiplicity of assumptions with the same shape is irrelevant.

2.3 Stoic Natural Deduction

Now that the *themata* have been dealt with, we can turn our attention to the indemonstrables. Indemonstrables are arguments which deduce propositions from propositions — a notion which, as noted by Bobzien in earlier works [4, pg. 182], naturally suggests their formalization as inferences of natural deduction.

The primitive kinds[7] of indemonstrables are defined as follows:

1. A first indemonstrable is an argument that from a conditional and its antecedent deduces the consequent of the conditional.

2. A second indemonstrable is an argument that from a conditional and the contradictory of the consequent deduces the contradictory of the antecedent of the conditional.

3. A third indemonstrable is an argument that from the negation of a conjunction and one of the conjuncts deduces the contradictory of the other conjunct.

4. A fourth indemonstrable is an argument that from a disjunction and one of the disjuncts deduces the contradictory of the other disjunct.

5. A fifth indemonstrable is an argument that from a disjunction and the contradictory of one of its disjuncts deduces the other disjunct.

And their formalization as inferences is straightforward:

$$\frac{A \quad A \to B}{B} \text{ A1} \qquad \frac{A \to B \quad B^*}{A^*} \text{ A2}$$

$$\frac{A_i \quad \neg(A_1 \wedge A_2)}{A_j^*} \text{ A3} \qquad \frac{A_i \quad A_1 \oplus A_2}{A_j^*} \text{ A4}$$

$$\frac{A_i^* \quad A_1 \oplus A_2}{A_j} \text{ A5}$$

---

[7]Using modern terminology, it seems accurate to say that the statements defining indemonstrables single out *types* of arguments, as they establish *argumentation schemas* for particular arguments instead of particular arguments themselves. We would like to thank one of the reviewers for this observation. However, it seems that the Stoics defined the indemonstrables themselves as particular arguments, not as *schemas*. Comments on this terminological aspect of Stoic texts can be found in [4, pgs. 134-135] and on footnote 24 of [3].

Restriction on A3, A4 and A5: $i, j \in \{1, 2\}$, $i \neq j$.

Curiously enough, all inferences/indemonstrables used in Stoic logic have the structure of elimination rules for their respective logical operators; only the first *thema* seems to have the structure of an introduction rule.

We can now finish the system and state our main definitions.

Definition 2.2. *Derivations of Stoic Natural Deduction (SND) are defined inductively as follows:*

1. A single instance of A1, A2, A3, A4 or A5 with premises A and B and conclusion C is a derivation of C from A and B;

2. If $\begin{array}{c}\Gamma\\ \Pi^1\\ A\end{array}$ is a derivation of A from Γ, Γ contains precisely two distinct formulae and $\begin{array}{cc}A & \Delta\\ \Pi^2 & \\ B & \end{array}$ is a derivation of B from $A \cup \Delta$,

   then $\begin{array}{cc}\Gamma & \\ \Pi^1 & \\ A & \Delta\\ \Pi^2 & \\ B & \end{array}$ is a derivation of B from $\Gamma \cup \Delta$;

3. If $\begin{array}{cc}\Gamma & A\\ \Pi & \\ B & \end{array}$ is a derivation of B from $\Gamma \cup A$, $\dfrac{\begin{array}{cc}\Gamma & [A]\\ \Pi & \\ B & B^*\end{array}}{A^*} T$

   is a derivation of A^* from $\Gamma \cup B^*$, provided that $B^* \notin \Gamma$.

Definition 2.3. $\Gamma \vdash_{SND} A$ *if and only if we can construct a derivation of A from Γ using the previous definition.*

The cardinality restriction imposed on the second clause is a peculiar feature of Stoic logic, and will be discussed in more depth at Section 2.4.

Equivalence between this system and Bobzien's Sequent Calculus follows immediatelly from the fact that the three *themata* can be substituted [2] by the following equivalent rule, with $A \neq B$:

$$\frac{A, B \Rightarrow C \quad C, \Delta \Rightarrow D}{A, B, \Delta \Rightarrow D} \, Cut^2$$

Remarkably, unlike the root-first proofsearch method of Bobzien's sequent calculus and the leaf-first derivation method of usual natural deduction systems, derivations on SND must be built from the *middle*, except if no applications of T are used. We start by writing down an indemonstrable, and then expand the derivation upwards by placing derivations on top of the premises. After any number of upward expansions, we may expand a derivation downwards by placing an instance of T at its end, and such procedures can be repeated as many times as needed. Even though the end result of this procedure looks very much like usual leaf-first forward derivations in natural deduction, there are derivations which can be built using the usual definitions but that are not valid deductions of SND. This is a direct consequence of the fact that Stoic logic does not satisfy *Cut* [2] and, therefore, does not admit unrestricted composition.

It is also important to note that, after applying an instance of T with premises A and A^*, we can further expand the derivation upwards to include either A or A^* on the set of dependencies, so that the restriction that premises of a T rule may not occur of the set of assumptions of the derivation becomes of little consequence. This seems to be Bobzien's original intent, as nothing in the Stoic's informal statement of the first *thema* corresponds to this restriction.

Just as in the case of sequent calculus, when we take rules A1 through A5 together with T, indemonstrables A6–A9 all hold as derivable rules, provided that the antecedents are distinct formulae (due to the restriction that the formula added by the instance of T cannot already be present on the derivation):

6. $A, B^* \vdash \neg(A \to B)$ (sequents analyzable into A1 or A2 axioms)

7. $A, B \vdash A \wedge B$ (sequents analyzable into A3 axioms)

8. $A, B \vdash \neg(A \oplus B)$ (sequents analyzable into A4 axioms)

9. $A^*, B^* \vdash \neg(A \oplus B)$ (sequents analyzable into A5 axioms)

Proofs:

$$\frac{A \quad [A \to B]}{B} \text{A1} \qquad \frac{B^*}{\neg(A \to B)} \text{T, } A \to B \qquad \frac{A \quad [\neg(A \wedge B)]}{B^*} \text{A3} \qquad \frac{B}{A \wedge B} \text{T, } \neg(A \wedge B)$$

$$\frac{A \quad [A \oplus B]}{B^*} \text{A4 L} \qquad \frac{B}{\neg(A \oplus B)} \text{T, } A \oplus B \qquad \frac{A^* \quad [A \oplus B]}{B} \text{A5 L} \qquad \frac{B^*}{\neg(A \oplus B)} \text{T, } A \oplus B$$

Those deductions make it clear that derivations of Stoic natural deduction are much more intuitive than their counterparts in Stoic sequent calculus. Consider, for instance, the proof of $p \to q, r \to q, p \oplus r \Rightarrow q$ presented by Bobzien in [3]:

$$\cfrac{\cfrac{r \to q, \neg q \Rightarrow \neg r}{} \text{A2} \quad \cfrac{p \oplus r, \neg r \Rightarrow p}{} \text{A5} \quad \cfrac{\cfrac{\overline{p, p \to q \Rightarrow q}}{p, \neg q \Rightarrow \neg(p \to q)} \text{T1}}{p \oplus r, \neg r, \neg q \Rightarrow \neg(p \to q)} \text{T3}}{\cfrac{r \to q, p \oplus r, \neg q \Rightarrow \neg(p \to q)}{p \to q, r \to q, p \oplus r \Rightarrow q} \text{T1}} \text{T4}$$

Its natural deduction counterpart would be the following deduction:

$$\cfrac{\cfrac{r \to q \quad [\neg q]}{\neg r} \text{A2} \quad p \oplus r}{\cfrac{\cfrac{p}{q} \text{A5} \quad [p \to q]}{\cfrac{\neg(p \to q)}{q} \text{T1, } \neg q} \text{A1} \quad [\neg q]} \text{T1, } p \to q \quad \cfrac{p \to q}{} \text{T1, } \neg q}$$

To give another example, consider this proof, provided in a footnote of [3]:

48

Stoic Natural Deduction

$$\cfrac{\neg p \to p, \neg p \Rightarrow p \quad A1 \qquad \cfrac{\cfrac{p, p \to p \Rightarrow p}{p, \neg p \Rightarrow \neg(p \to p)}\ A1}{p, \neg p \Rightarrow \neg(p \to p)}\ T1}{\cfrac{\neg p \to p, \neg p \Rightarrow \neg(p \to p)}{\neg p \to p, p \to p \Rightarrow p}\ T1}$$

Its natural deduction counterpart would be the following:

$$\cfrac{\cfrac{[\neg p] \quad \neg p \to p}{p}\ A1 \quad [p \to p]}{\cfrac{\neg(p \to p)}{p}\ T1, p \to p} \quad \cfrac{p \to p}{}\ T1, \neg p$$

2.4 Stoic Natural Deduction with unrestricted compositionality

The main difficulty faced by definitions of leaf-first Stoic deduction is the restricted version of *Cut* used in Bobzien's calculus, which makes it necessary to include extraneous restrictions on deductions to produce similar effects. For instance, if we have a derivation of A which ends with and application of T and has three or more open assumptions, even though it can be further expanded downwards by more applications of T, it cannot be expanded by an application of A1–A5 which uses the conclusion of the derivation as one of the premises. This is what happens in Bobzien and Dyckhoff's counterexample for the admissibility of cut in Stoic logic:

$$\cfrac{p \to q, p \to (q \to r), \neg r \Rightarrow \neg p \qquad \neg p, \neg p \to s, \neg p \to (s \to t) \Rightarrow t}{p \to q, p \to (q \to r), \neg r, \neg p \to s, \neg p \to (s \to t) \Rightarrow t}\ \text{Cut}$$

The sequents above the rule are derivable, but the sequent below it is not. In SND, we can easily write derivations which correspond to the upper sequents:

49

$$\dfrac{[p] \quad p \to q}{q} \qquad \dfrac{\dfrac{[p] \quad p \to (q \to r)}{q \to r}}{\dfrac{r}{\neg p} \qquad \neg r}$$

$$\dfrac{\neg p \quad \neg p \to s}{s} \qquad \dfrac{\neg p \quad \dfrac{\neg p \quad \neg p \to (s \to t)}{s \to t}}{t}$$

We cannot, however, join the derivations by putting the first one above the instances of $\neg p$ on the second. If we begin constructing the derivation from the instance of A1 which concludes r from q and $q \to r$, no part of the definition of what counts as a derivation will allow us to appropriately expand it downwards after obtaining $\neg p$, as downward expansions can only be done via applications of T. If, on the other hand, we begin constructing the derivation from the instance of A1 which concludes t from s and $s \to t$, we will not be allowed to expand it upwards from $\neg p$ due to the fact that the first derivation has three undischarged premises.

But why does Stoic logic place such strict cardinality constraints on composition? Even though the Stoic requirement that derivations must have at least two undischarged premises may be philosophically justified [4, pgs. 171-180], there seems to be no practical reason to deny that the composition of those two derivations would yield a valid derivation. In fact, the relevant literature seems to contain no reports of Stoic arguments in favour of restrictions of this kind, and the composition of those derivations do not fit in the classification of invalid arguments given by the Stoics[8].

It is possible, then, that the Stoics wished to define a set of rules jointly equivalent to unrestricted *Cut* rules, but ended up accidentally defining a strictly weaker set of rules. This becomes even more plausible if we consider that the rules seem equivalent at first glance, both to ancient and contemporary logicians. According to Bobzien, Alexander of Aphrodisias stated that the Stoic *themata* were obtained by simplifying the "synthetic theorem" of the peripatetics (a claim regarded as historically unlikely [2]), and then mistakenly asserted that the three

---

[8] For a brief exposition of this classification, see [10, pg. 164–166].

themata were jointly equivalent to the synthetic theorem (which is, in turn, equivalent to unrestricted *Cut*). Furthermore, even Bobzien herself speculated in earlier works [4, pg. 167] that multiplicity of antecedents would be redundant and the synthetic theorem would have no more deductive power than the *themata*, something that was disproved only much later by Bobzien and Dyckhoff in [2].

We have no way of knowing if the Stoics were aware of those technical limitations and, if they weren't, how they would respond to them. But, as long as one is more interested in the further advancement of Stoic-like systems than in providing faithful historical reconstruction of the original logic (which is sufficiently covered by Bobzien's many works), there seems to be strong reasons for adopting frameworks which allow unrestricted *Cut* rules and compositions.

Definition 2.4. *The system of Unrestricted Stoic Natural Deduction (SUND)*[9] *is obtained by taking together rules T, A1, A2, A3, A4 and A5 together with the usual definition of derivations for natural deduction*[10]*, but with the following additional restrictions:*

1. *Derivations must have at least length 3*[11]*;*

2. *At every step of the derivation, the conclusion must depend on at least two distinct undischarged assumptions;*

3. *The conclusion A depends on the set Γ of assumptions only if all assumptions of Γ are effectively used*[12] *on the derivation.*

Definition 2.5. $\Gamma \vdash A$ *is valid whenever there is a derivation of A from Γ in SUND.*

---

[9]Even though it does not follow the phrase it stands for, the name SUND was adopted because it is easier to pronounce than USND. A similar convention will also be adopted with the intuitionistic system ISUND.

[10]See, for example, the definitions of [19] or [6].

[11]That is, they must always contain at least three formula occurrences [19, pg. 26], which amounts to saying that we start derivations by writing down an indemonstrable instead of a single formula occurrence.

[12]That is, all formulas in Γ must actually appear on the derivation, in contrast to definitions on which $\Gamma \vdash A$ holds whenever there is a derivation of A on which a *subset* of Γ appears on the derivation.

Alternatively, one could obtain SUND by adapting Definition 2.2., which would again yield a fruitful proofsearch method. The first restriction is made in order to avoid derivations of length one such as $A \vdash A$, on which A is a single assumption derived from itself with no rules. The second restriction is made in order to preserve the structure of Stoic sequents, preventing derivations such as $A, A \vdash A \wedge A$ (for why this is desirable in Stoic systems, see [3]). The third restriction is made in order to preserve relevance, as the usual definitions of deduction regard the presence of a proposition A on any derivation as a sufficient, but not necessary, condition for including it in the set Γ of formulae on which it depends.

It is also worthy of note that being easily constructible both from a root-first and from a leaf-first perspective is another desirable property of Stoic systems — for, as stated by Gould, "the Stoics appear to have set forth both a discursive method of demonstration and a test for validity"[10]. SND is indeed well-behaved from a root-first perspective (test of validity), but leaves much to be the desired from a leaf-first perspective (method of demonstration). We argue, then, that SUND retains this root-first good behavior whilst significantly improving the systems's leaf-first functionality, allowing it to be more easily used in demonstrations.

2.5 Intuitionistic Stoic Logic

New variants of the Stoic system may also provide interesting solutions for longstanding problems of logic. For instance, a discussion between Johansson and Heyting [17] about what kind of behavior should be expected from an intuitionistic formulation of the conditional would seem to benefit from some particular features of Stoic logic. Johansson's original goal [12] was to remove the following two axioms from Heyting's [11] intuitionistic logic:

1. $B \to (A \to B)$

2. $A \to (\neg A \to B)$

Logics that require a special relevance relation to hold between two formulae A and B in order for a proposition $A \to B$ to be asserted are called *relevance logics*, and logics that prevent us from concluding arbitrary formulae from contradictions are called *paraconsistent logics*. Axiom 1 is regarded as problematic by relevance logicians because, since A is an arbitrary proposition (which may thus be validly substituted by any other), the truth of B entails the truth of $A \to B$ regardless of which proposition A we choose and of any relevance relation between A and B. Axiom 2 is regarded as problematic by paraconsistent logicians because B is an arbitrary proposition concluded from the contradictory propositions A and $\neg A$[13], and also by relevance logicians because the truth of A entails the truth of $\neg A \to B$ for any B regardless of any relevance relation between $\neg A$ and B. The remaining axioms of intuitionistic logic are usually regarded as unproblematic by both traditions. Only the second axiom was absent in the final version of Johansson's minimal logic, which makes it so that minimal logic is a *paraconsistent logic*, but not a *relevance logic*.

The non-derivability of both axioms in Stoic logic follows trivially from the fact that Stoic logic has no theorems. But, more importantly, the *inferences* those axioms represent are also not valid in Stoic logic — which suggests that Stoic Logic is both a *relevance logic* and a *paraconsistent logic*. Since there is no implication introduction rule, a Stoic derivation $\Gamma \vdash B$ cannot in general be expanded to a derivation of $\Gamma \vdash A \to B$ and, due to the absence of any rule equivalent to the principle of explosion, we cannot use two derivations $\Gamma \vdash A$ and $\Gamma' \vdash \neg A$ to conclude $\Gamma, \Gamma' \vdash B$ for arbitrary B[14].

To see why this is so, consider an arbitrary derivation $\Gamma \vdash B$. If we apply an indemonstrable to the end of the derivation, the conclusion will either be a subformula of B or the contradictory of a subformula of B, and as such cannot be $A \to B$. The only possibility, then, is to apply an instance of T. But we can only apply T without adding premises to Γ

---

[13]The arbitrary character of the conclusion is essential to characterize a paraconsistent logic. Minimal logic is a paraconsistent logic because from a proposition A and its negation $\neg A$ we cannot conclude an arbitrary proposition B, but it does validate a weakened, "negative" version of the *ex falso*: from the propositions A and $\neg A$ we can derive $\neg B$, for arbitrary B.

[14]Notice that this would not be the case if applications of T without effective discharge were allowed, since then it could be shown that the explosion principle holds in the same way that $\bot_i$ can be seen as a special case of $\bot_c$ [19, pg. 21].

if the contradictory of the application's conclusion is already in Γ (due to the restriction of effective discharge) and there is a set Δ such that $\Delta \subseteq \Gamma$, $\neg(A \to B) \in \Delta$ and $\Delta \vdash B^*$, so that we can join $\Gamma \vdash B$ and $\Delta \vdash B^*$ to discharge the formula $\neg(A \to B)$ in Γ with an application of T and conclude $\Gamma \vdash A \to B$, since $((\Gamma - \{\neg(A \to B)\}) \cup \Delta) = \Gamma$. Those conditions are highly specific and do not hold for most sets[15], which shows that the inference is not in general valid.

For the second inference, let $\Gamma \vdash A$ and $\Gamma' \vdash A^*$. The only rule which could join those derivations to produce an arbitrary conclusion B is T. However, since discharge is mandatory, the contradictory of B must already be in either Γ or Γ'. Since there is an infinite number of possible propositions, derivations are finite and $\Gamma \vdash A$ is valid only when there is a derivation of A which effectively depends on Γ, if we let B be any proposition whose contradictories do not occur on the assumptions of Γ or Γ', it follows immediatelly that $\Gamma, \Gamma' \nvdash B$.

In order to use *reduction ad absurdum* in Stoic logic, we must rely on the definition of contradictory propositions, which directly introduces double negation elimination on the system. This notion can be used either together with the T rule to produce a derivation which discharges $\neg A$ and concludes A or with indemonstrables which use it to indirectly eliminate double negations (e.g. an instance of A2 which concludes A from $\neg A \to B$ and $\neg B$).

Thus, in order to obtain an intuitionistic version of Stoic logic, it suffices to either remove the notion of contradictory propositions or adapt it in such a way as to prevent $\neg\neg A$ and A from being considered equivalent. The most economic adaptation is one in which we simply abandon Definition 2.1. and substitute all formulae A^*, B^* and C^* occurring on the definitions of T and A1–A5 by $\neg A$, $\neg B$ and $\neg C$, but a new variation of T must also be added to account for a subtle loss of expressive power in T. Furthermore, in order to obtain a complete system, it is also prudent to substitute the primitive indemonstrable A3 by the primitive indemonstrable A7; in intuitionistic Stoic logic, A3 can be analyzed into A7, but A7 cannot be analyzed into A3.

We end up with the following system:

---

[15] In fact, it seems that there cannot be such a set Δ, but we do not have a proof of this yet.

$$\frac{A \quad A \to B}{B} \text{A1} \qquad \frac{A \to B \quad \neg B}{\neg A} \text{A2}$$

$$\frac{A_i \quad A_j}{A_i \wedge A_j} \text{A7} \qquad \frac{A_i \quad A_1 \oplus A_2}{\neg A_j} \text{A4}$$

$$\frac{\neg A_i \quad A_1 \oplus A_2}{A_j} \text{A5}$$

$$\frac{\Gamma \quad [A]}{\vdots} \\ \frac{C \quad \neg C}{\neg A} \text{T1}, A \qquad \frac{\Gamma \quad [A]}{\vdots} \\ \frac{\neg C \quad C}{\neg A} \text{T2}, A$$

Restriction on A7, A4 and A5: $i, j \in \{1, 2\}$, $i \neq j$.

Restrictions on T1 and T2: there must be at least one assumption with shape A above $C/\neg C$, and discharge of all assumptions with such shape is mandatory.

Definition 2.6. *The system of Intuitionistic Unrestricted Stoic Natural Deduction (ISUND) is obtained by taking together rules T1, T2, A1, A2, A7, A4 and A5 together with the usual definition of derivations for natural deduction, but with the following additional restrictions:*

1. *Derivations must have at least lenght 3;*

2. *At every step of the derivation, the conclusion must depend on at least two distinct undischarged assumptions;*

3. *The conclusion A depends on the set Γ of assumptions only if all assumptions of Γ are effectively used on the derivation.*

Definition 2.7. *$\Gamma \vdash_i A$ is valid whenever there is a derivation of A from Γ in ISUND.*

We could also, of course, obtain an intuitionistic version of SND by adapting definition 2.2., which would yield an intuitionistic variant with the same restrictions on compositionality.

As noted in [2], all inferences of Stoic logic are classically valid, a result which can easily be extended to SUND (using, for example, the natural deduction rules defined in [19] to show that all Stoic inferences are derivable, and defining exclusive disjunction in terms of other connectives). The same result can be proven for ISUND and intuitionistic logic. For A1, A2 and A7 the proof is trivial, but in the case of A4 and A5 we must make adaptations due to the fact that exclusive disjunction is not usually defined in constructive systems. By letting $A \oplus B := (A \wedge \neg B) \vee (B \wedge \neg A)$, we can use derivations of the following kind:

$$\cfrac{(A \wedge \neg B) \vee (B \wedge \neg A) \quad \cfrac{[A \wedge \neg B]}{A} \quad \cfrac{\cfrac{[B \wedge \neg A]}{B} \quad \neg B}{\cfrac{\bot}{A}}}{A} \; A \wedge \neg B, B \wedge \neg A$$

$$\cfrac{(A \wedge \neg B) \vee (B \wedge \neg A) \quad \cfrac{[A \wedge \neg B]}{\neg B} \quad \cfrac{\cfrac{[B \wedge \neg A]}{\neg A} \quad A}{\cfrac{\bot}{\neg B}}}{\neg B} \; A \wedge \neg B, B \wedge \neg A$$

The proofs for the remaining instances of A4 and A5 are similar.

One could possibly object that defining $\oplus$ in terms of $\wedge$, $\vee$ and $\neg$ is inadequate, as intuitionistic operators are expected to be independent from each other. In the case of exclusive disjunction, however, this is almost inescapable; a connective which demands either a proof of A and of B but does not allow simultaneous proofs of A and B easily collapses into the remaining connectives.

Consider, for example, the following rules:

$$\frac{A \quad \overset{[B]}{\underset{\bot}{\vdots}}}{A \oplus B} \text{ I} \oplus_1 \qquad \frac{B \quad \overset{[A]}{\underset{\bot}{\vdots}}}{A \oplus B} \text{ I} \oplus_2$$

$$\frac{A \oplus B \quad A \quad B}{\bot} \text{ E} \oplus_1 \qquad \frac{A \oplus B \quad \overset{[A]}{\underset{C}{\vdots}} \quad \overset{[B]}{\underset{C}{\vdots}}}{C} \text{ E} \oplus_2$$

Even though they seem to encapsulate what we mean by a exclusive disjunction, the new operator immediatelly collapses into the formula which we used for the definition:

$$\frac{\dfrac{A \quad \dfrac{\overset{[B]}{\vdots}}{\neg B} \quad B}{A \wedge \neg B}}{(A \wedge \neg B) \vee (B \wedge \neg A)} \qquad \frac{\dfrac{B \quad \dfrac{\overset{[A]}{\vdots}}{\neg A} \quad A}{B \wedge \neg A}}{(A \wedge \neg B) \vee (B \wedge \neg A)}$$

$$\frac{(A \wedge \neg B) \vee (B \wedge \neg A) \quad \dfrac{\dfrac{[A \wedge \neg B]}{\neg B} \quad B}{\bot} \quad \dfrac{\dfrac{[B \wedge \neg A]}{\neg A} \quad A}{\bot}}{\bot} \; A \wedge \neg B, B \wedge \neg A$$

$$\frac{(A \wedge \neg B) \vee (B \wedge \neg A) \quad \dfrac{[A \wedge \neg B]}{\underset{C}{\overset{A}{\vdots}}} \quad \dfrac{[B \wedge \neg A]}{\underset{C}{\overset{B}{\vdots}}}}{C} \; A \wedge \neg B, B \wedge \neg A$$

Therefore, it seems reasonable that the definition of $\oplus$ is not independent from the remaining connectives, even when defined in intuitionistic logic.

To conclude this section, we point out that ISUND is a proper subsystem of intuitionistic logic, but *not* of minimal logic. Disjunctive syllogism fails in minimal logic due to the absence of $\bot_i$, and we use $\bot_i$ in an essential way to show that $A \oplus B, \neg B \vdash A$. As such, provided we define the exclusive disjunction as above, it seems that ISUND is weaker than intuitionistic logic and, in many respects, also than minimal logic, but that its disjunction is stronger than that of minimal logic — as it allows a limited, implicit use of *ex falso* to validate the disjunctive syllogism.

Even though there are critics of the disjunctive syllogism[16], it is usually regarded as a desirable inference, but also one that requires the abandonment of either relevance or paraconsistency[17]. Intuitionistic Stoic Logic can thus be regarded as one of the few *paraconsistent, relevant* logics which allows the use of disjunctive syllogism.

3 Concluding remarks

Stoic logic has some very interesting features from the point of view of contemporary formal logic. Aside from being relevant and paraconsistent, it is a system finely tailored for demonstrations of logical consequence in a very strict sense.

While in usual natural deduction we define introduction rules for operators and obtain their elimination rules as a consequence of the rule's inversion [19, pg. 33], Stoic logic uses only elimination rules, and so defines its operators exclusively in terms of what can be inferred from them. Just like the Stoic's classification of the first *thema* defies Gentzen's division of rules into operational and structural, the Stoic's definition for the indemonstrables defy Prawitz's [19] and Dummet's [7] meaning explanation for logical connectives.

Stoic inferences typically takes us to a conclusion which is either strictly simpler than the premises or the contradictory of a proposition strictly simpler than the premises[18]. We are often only allowed to either

---

[16] See, for example, [20].

[17] With some noteworthy exceptions; see, for example, the systems presented in [21].

[18] Sadly, rule A7 on Intuitionistic Stoic Logic is an exception, but one can always substitute it for A3 to bring back this property — at the cost of abandoning A7 as a valid inference altogether.

conclude propositions which were already *contained* in the premises or to get rid of inconsistencies in our original presuppositions. Naturally, since no propositions are contained on the empty set, no consequences can be derived from an empty set of premises. Thus, Stoic logic allows us to reason about *truths contained in truths we have already assumed*

This restricted but useful notion of logical consequence makes it extremely simple to deal with problems such as the paradoxes of material implication, and the solutions provided by the system have not lost their elegance with time. In light of all this, we can only hope that Stoic logic will soon receive the attention it deserves from logicians, as well as hope that Stoic proof theory is developed even further after this short paper.

References

[1] R. Barcan Marcus, 'Modalities and Intensional Languages' in *Synthese*, 1961, 13 (4), pp. 303–322.

[2] S. Bobzien and R. Dyckhoff, 'Analyticity, Balance and Non-admissibility of Cut in Stoic Logic' in *Studia Logica*, 2019, 107, pp. 375–397.

[3] S. Bobzien, 'Stoic Sequent Logic and Proof Theory' in *History and Philosophy of Logic*, 2019, 40 (3), pp. 234–265.

[4] S. Bobzien, 'Stoic Syllogistic' in *Oxford Studies in Ancient Philosophy*, 1996, 14, pp. 92–133.

[5] D. Bostock, *Intermediate Logic*. Oxford University Press, Oxford, 1997.

[6] D. van Dalen, *Logic and Structure*. Springer-Verlag, London, 2013.

[7] M. Dummett, *The Logical Basis of Metaphysics*. Harvard University Press, Cambridge, 1991.

[8] G. Frege, 'Begriffsschrift' in J. Van Heijenoort (ed.) *From Frege to Gödel*. Harvard University Press, Cambridge, 1967, pp. 1–83.

[9] G. Gentzen, 'Investigations into Logical Deduction' in M. E. Szabó (ed.) *The Collected Papers of Gerhard Gentzen*. North-Holland Publishing Company, Amsterdam, 1970, pp. 68–131.

[10] J. Gould, 'Deduction in Stoic Logic' in J. Corcoran (ed.) *Ancient Logic and Its Modern Interpretations*. Springer Netherlands, Dordrecht, 1974, pp. 151–168.

[11] A. Heyting, *Intuitionism: An Introduction*. North-Holland Publishing Company, Amsterdam, 1971.´

[12] I. Johansson, 'Der Minimalkalkul, ein Reduzierter Intutionischer Formalismus' in *Compositio Mathematica*, 1937, 4, pp. 119-136.

[13] S. Kripke, 'A Completeness Theorem in Modal Logic' in *Journal of Symbolic Logic*, 1959, 24 (1), pp. 1-14.

[14] C. I. Lewis, 'The Calculus of Strict Implication' in *Mind*, 1914, 23 (90), pp. 240-247.

[15] J. Lukasiewicz, *Aristotle's Syllogistic from the Standpoint of Modern Formal Logic*. Oxford University Press, Oxford, 1957.

[16] B. Mates, *Stoic Logic*. University of California Press, Berkeley and Los Angeles, 1961.

[17] T. van der Molen, *The Johansson/Heyting Letters and the Birth of Minimal Logic*. Technical Report X-2016-04, ILLC Publications, Amsterdam, 2016.

[18] J. von Plato, *Elements of Logical Reasoning*. Cambridge University Press, Cambridge, 2014.

[19] D. Prawitz, *Natural Deduction: A Proof-theoretical Study*. Dover Publications, New York, 2006.

[20] S. Read, 'What Is Wrong with Disjunctive Syllogism?' in *Analysis*, 1981, 41 (2), pp. 66–70.

[21] N. Tennant, *Core Logic*. Oxford University Press, Oxford, 2017.

Seneca's and Porphyry's Trees in Modern Interpretation

Jens Lemanski
University of Münster, University of Hagen, Germany
`jens.lemanski@fernuni-hagen.de`

1 Introduction

Traditional tree diagrams, as developed in Platonic and Aristotelian doctrine (see e.g. Fig. 1) are regarded as precursors of modern techniques of visualisation in philosophy, biology, mathematics, linguistics, computer science, music theory etc. [16], [44], [10, chap. 5]. These visualisations are used to constitute ontologies, taxonomies or, generally speaking, to perform conceptual analysis [17, p. xiii]. If we take only computer science as an example, these tree diagrams have been used in areas such as ontology engineering, semantic web, object-oriented programming, knowledge representation, artificial intelligence, etc. [8]. John Sowa, for example, wrote that the first semantic network in the form of a tree diagram can be found in the work of the Neoplatonist Porphyry (c. 234–305 AD) [41, p. 4]. This assessment was also reproduced in *AIMA*, today's standard textbook on artificial intelligence [3, p. 471]. Nowadays, we also know that perhaps the first design of a logic machine in the Baroque era was inspired by the form of tree diagrams [30, p. 10], [Sect. 5].

The following information often reappear in connection with tree diagrams: Logicians and metaphysicians in earlier times generally (1) used only one (porphyrian) syntax of tree diagram, (2) usually made

---

Parts of this paper were presented at a lecture on 14 January 2021 at Sun-Yat Sen University and on 8 October 2021 during the conference 'Concepts of Logic, Logic of Concepts' at the University of Leipzig. I would like to thank the discussion partners and especially Ingolf Max for many valuable suggestions. The paper benefited from the project 'History of Logic Diagrams in Kantianism' (Thyssen Stiftung).

no nominal distinction between different types of trees and (3) almost always illustrated one semantics, namely the concepts depending on the Aristotelian category of **substance** as the highest genus in the tree diagram. However, many of these prejudices have been revised in detailed studies: (1) Barnes [29, p. 108ff.] Mansfeld [25, p. 78ff.] and further also Verboon [46, pp. 44ff.] argue, for example, that one should distinguish between at least two syntactic forms of traditional tree diagrams. (2) Barnes and Mansfeld also argue that these two diagram types can be traced back to two classical philosophical texts, i.e. Seneca's *58th letter to Lucilius* and Porphyry's *Isagoge*. (3) Blum [5, p. 301] and Sowa [41] note that tree diagrams can be used to depict more categories than just the one of **substance**.

My original aim was to support all these three points from the aforementioned studies. I had the idea to focus on many unusual diagrams from the scholastic and early modern periods and to show that they depict and represent much more information than one would expect. In doing so, however, I had to realise that a modern interpretation of the classical texts and diagrams, which are connected to so-called Seneca's and Porphyry's trees, is so wide-ranging that one does not get to advance from antiquity to modern times in the scope of an ordinary paper.

The most serious problem I have seen is the way traditional tree diagrams have been treated in the literature: While there is a growing body of historical work on tree diagrams in the medieval and early modern periods, none offers a – from a logical standpoint – satisfactory way of describing them: The approaches of Sommers and Englebretsen are profitable for term logic, but perhaps only serve to a limited extent for the analysis of traditional tree diagrams [12]. There are many good logical approaches from the field of formal ontology, but they often only examine individual sub-questions of tree diagrams [15]. Hacking [16], who is one of the few to combine a historical and logical approach to tree diagrams, sees a continuous development from antiquity to modern graph theory, but he too only outlines some difficulties in applying modern graph theory to traditional tree diagrams. But that is exactly what one needs to look at more complex tree diagrams, which have been around since the early Middle Ages.

For this reason, I would like to argue in this paper for a modern interpretation of traditional tree diagrams, continuing mainly the approach of Hacking. In Section 2, I will first introduce the two classical texts by Seneca and Porphyry. Section 3 draws on the interpretations of Barnes and Mansfeld, who have the most convincing approach to explain how the texts of Seneca and Porphyry are transformed to tree diagrams from the early Middle Ages onwards. In Section 4, I will then discuss the semantics and syntax of traditional Seneca and Porphyry trees. Section 5 will then sketch some problems and examples of the given syntax and semantics. Finally, Section 6 will give a summary and an outlook.

2 Trees in Seneca and Porphyry

If the intention is to examine ancient texts that could be a trigger for the great flood of tree diagrams that have come down to us from the early Middle Ages at the latest, one must actually look into the sources before Seneca and Poryphyry. Plato and Aristotle are relevant authors to whom later philosophers such as Seneca and Porphyry explicitly referred. Certainly, however, traces can already be found in the Pre-Socratics. Nevertheless, the texts of Seneca and Porphyry are considered seminal for the visualisation of conceptual structures, which then became known as tree diagrams. In this respect, it makes sense to examine not the entire genealogy of tree diagrams, but the most important passages of the texts.

In this section, I would first like to discuss the relevant text passages by Seneca (2.1) and Porphyry (2.2). As will be shown in Section 3, both text passages are the foundations for later tree diagrams due to their metaphorical way of speaking, but do not contain any visualisations themselves.

2.1 Seneca's *Letter 58*

Seneca's 58th letter has two central themes, especially from the perspective of logic, i.e. Platonic concepts of οὐσία (essentia) and τὸ ὄν (quod est) [48, p. 622ff.] and the relationship between *genus* and *species*. In Seneca's treatise on genus and species two different methods are involved: (1) perductio and (2) deductio.

(1) In perductio, singuli or single items are picked up backwards (*coeperimus singula retro legere*). By collecting and connecting more and more singuli higher and higher species and genera emerge bottom-up. The fact that we are picking up the singuli backwards indicates that there was already a forward movement that distributed these singuli. The method of perductio, which Seneca unfortunately does not explain in detail, is particularly reminiscent of *inductio by simple enumeration*, but also to some extent of the processes that today are called *backward chaining* [6, Chap. 13.2], [28].

(2) Once Seneca has arrived at the highest genus, i.e. the `being` or `quod est`, he deduces all subsequent subconcepts top-down with the help of the division [2, p. 223], [48, I, p. 98f.]. In doing so, Seneca uses three theoretical terms that are intended to structure a set of concepts that depend on the Platonic term `quod est`. In order to make the theoretical terms clear in the following, I insert the Latin expressions in italics in curly brackets and use them in the following:

> For by using this term [sc. `quod est`] they will be divided into species, so that we can say: that which exists either possesses, or lacks, substance. This, therefore, is what genus is, — the primary, original, and (to play upon the word) 'general' {*genus generale*}. Of course there are the other genera: but they are 'special' genera {*genera specialia*}: 'man' being, for example, a genus. For 'man' comprises species: by nations, — Greek, Roman, Parthian; by colours, — white, black, yellow. The term comprises individuals {*singuli*} also: Cato, Cicero, Lucretius. So 'man' falls into the category genus, in so far as it includes many kinds; but in so far as it is subordinate to another term, it falls into the category species. But the genus 'that which exists' [sc. `quod est`] is general {*genus generale*}, and has no term superior to it. It is the first term in the classification of things, and all things are included under it. [37, p. 393ff.]

The quoted passage, which deals with the division of the genus `being` (`quod est`) into several species and singuli, comes from the part on deductio, in which Seneca proceeds top-down. In this quote, Seneca introduces three theoretical terms: (1) The *genus generale* is the `quod est` and stands highest. (2) Below the genus generale are many *genera specialia*, which are both species for higher genera and genus for lower species. (3) At the bottom are the *singuli*, which are contained by only one particular species. Each genus is usually divided into two subspecies or -concepts (dichotomic), sometimes in three or more (polytomic) using the divisio method.

To make it easier to assign the concepts to the corresponding technical terms, I have created Table 1. The terms in Tab. 1 are only a selection of the concepts mentioned in Seneca's text and are mainly oriented towards Mansfeld's selection and interpretation, which plays an important role in Sect. 3. In the last column, '1' indicates that it is true that a concept is a genus or species for something else; '0' indicates the opposite, i.e. that it is false or not the case that a concept is a genus or species for something else. As we can see, combinatorics of '1' and '0' is not exhaustive.

Although Seneca does not describe or draw a tree in the treatise on genera and species, there are metaphors of subordination that lend meaning to the text only through their arrangement in a vertical scheme (suspensa; sub se habere; superiorem locum; superius; supra se habet; sub illos; etc.): In the method called perductio, the text describes the ascent from the singuli to the genus generale. In the method of divisio, the descent from the genus generale to the singuli is given. The vertical image of the bottom-up or top-down movement can evoke the picture of the trunk of a tree, the diviso from one to many (or the perductio from many to one) the respective branches.

What is astonishing about Seneca's text is that he is very imprecise in his choice of terms, as some are nouns (e.g. `animal`, `horse`), others adjectives (e.g. `corporeal`, `animate`). But he is very precise in his use of logical connectives: He often uses the logical connective *exclusive or* to relate to a dichotomic pair, suggesting an oppositional relationship between a positive concept and its negation. Seneca uses the phrase *either...or*, i.e. *aut...aut*, aptly in a total of five passages and also explicitly refers in this context to the law of excluded middle ('Nihil tertium est.'). However, in the passages where Seneca subsumes three or more concepts under one generic one, he uses the metaphor of subordination

65

technical term	concepts	is genus/species for sth.
genus generale	quod est	1/0
genera specialia	corporeal/ incorporeal, animate/ inanimate, animal/ plant, man/ horse/ dog, Greeks/ Romans/ Parthians	1/1
singuli	Cato, Cicero, Lucretius	0/0

Table 1: Seneca's Technical Terms

or containment (comprehensa sunt, complectatur, in se, continet, etc.). The above quotation proves these relations.

Thus one finds in Seneca a meaningful connection of four themes that are again being discussed intensively and in context in logic today: classical negation, contradictory, dichotomy/ polytomy, and the laws of thought [4], [36]. Seneca thus seems to have been much better aware of logical relations than most modern historians of logic give him credit for.

2.2 Porphyry's *Isagoge*

Porphyry's relevant text, which later became known under the title *Introduction* (εἰσαγωγή), is a letter to a student named Chrysaorius. Porphyry explains at the beginning of the letter that he would like to introduce five terms in order to present a concise exposition to the Aristotelian *Organon* in the manner of an introduction (ὥσπερ ἐν εἰσαγωγῆς τρότῳ). It is usually thought that Porphyry only wanted to give an introduction to the first book of the *Organon*, namely the treatise on the concepts or *Categories*. However, since he also lists topics concerning the doctrine of judgement (τῶν ὁρισμῶν, τά περὶ διαιρέσεως) and inference (καί ἀποδείξεως), his introduction is not limited to the categories: Porphyry also has in mind the other books that are traditionally also counted as part of the Aristotelian *Organon*. Moreover, Porphyry indicates that the introduction is a summary of the knowledge of the ancients (πρεσβυτέροις), by which Seneca, among others, may be meant.

To this end, he explains five central concepts that became canonical from the Middle Ages onwards as *quinque voces* or *predicabilia*, i.e. genus, species, difference, property, and accident. Porphyry explains that he does not intend a metaphysical treatise but a logical one (λογικώτερον), as he omits topics of metaphysics and focuses mainly on the logical relations of the predicabiliae. The text can be divided into five chapters, each corresponding to one of the praedicabilia. In the first chapter. i.e. on genera and species, he defines four theoretical terms that can be distinguished by their combinatorial relation to genus and species. In order to make the theoretical terms clear in the following, I insert the Latin expressions (which became common by the translation of Boethius) in italics in curly brackets and use them in the following:

> For of predicates, some are said of only one item—namely, individuals {*individua*} (for example, Socrates and 'this' and 'that'), and some of several items—namely, genera and species [...]. In each type of predication there are some most general items {*genus generallissimum*} and again other most special items {*species specialissima*}; and there are other items between {*inter alia*} the most general and the most special. Most general is that above which there will be no other superordinate genus; most special, that after which there will be no other subordinate species; and between the most general and the most special are other items which are at the same time both genera and species (but taken in relation now to one thing and now to another). [29, p. 4–6]

The four theoretical terms can be distinguished by the extent to which they are genus or species for something else. (1) The *genus generallissimum* is genus for all other concepts, but it is not itself a species in relation to a higher concept. (2) The *inter alia* are both genus for some concepts and species for other ones. (3) The *species speciallissima* are species for other concepts, but not genus for any other one. (4) The *individua* are neither species nor genus for other concepts. Porphyry introduces the technical term *species speciallissima* that did not exist in Seneca. This is already made evident by the fact that Porphyry exhausts the possible combinations of genus and species, which was still incomplete in Seneca. To make the combinatorics of the theoretical terms even clearer, their relationship can be tabulated. I follow the same method as in Sect. 2.1.

Table 2 shows not only the four technical terms and the extent to which they are genus or species, but also which concepts are assigned to the four technical terms. The assignment goes back to another passage in the text, which was decisive for later interpreters and commentators in constructing a Porphyrian tree in the first place. Like Seneca, Porphyry does not visualise a tree, but only evokes a vertical scheme of concepts with his figurative terminology. In Porphyry, too, one finds a strong use of metaphors of subordination (e.g. ὑπὸ τὸ γένος) and containment (e.g. περίεχειν). In the relevant passage, Porphyry explains the four technical terms and their relationship to genus and species on an Aristotelian category:

> What I mean should become clear in the case of a single type of predication. Substance is itself a genus. Under it is body, and under body animate body, under which is animal; under animal is rational animal, under which is man; and under man are Socrates and Plato and particular men. Of these items, substance is the most general and is only a genus, while man is the most special and is only a species. Body is a species of substance and a genus of animate body. Animate body is a species of body and a genus of animal. Again, animal is a species of animate body and a genus of rational animal. Rational animal is a species of animal and a genus of man. Man is a species of rational animal, but not a genus of particular men—only a species. [29, p. 6]

Barnes rightly claims from this passage that the concepts mentioned in Tab. 2 would evoke a subsuming line rather than the idea of a tree [29, p. 109]. It is only in the second chapter on difference (διαφορὰ, lat. differentia) that another method appears that suggests a dichotomous division of concepts [29, p. 177ff.]: **Substance** can e.g. be divided by this dichotomy into **corporeal** and **incorporeal** or **body** into **animate** or **inanimate** etc. This results in a strict division between nouns and adjectives, sometimes called the division between extensional and intentional terms, since extension means the set of objects contained under a

technical terms	concepts	is genus/species for sth.
genus generalissimum	substance	1/0
inter alia	body, animate body, animal, rational animal	1/1
species specialissima	man	0/1
individua	Socrates, Plato, etc.	0/0

Table 2: Porphyry's Technical Terms

noun and intension means the properties contained under an adjective [15, p. 540],[1, pp. 45ff].

It is striking that these adjectives brought about by division are classical negations of each other: `rational` is the negation of `irrational`, `mortal` of `immortal`, etc. One can imagine the vertically arranged line of nouns as the trunk of a tree, and the dichotomously ordered adjectives as the branches that descend from this trunk. We will analyse this in more detail in Section 5.

Porphyry describes the relation of these adjectives with the expression of dihairetic or divisive difference (διαιρετικαὶ διαφοραί, lat. divisivae differentiae). If two adjectives stem from one noun for which it is true that it is a genus for another, then the positive adjectives describes a property of the next lower noun specie, e.g. `animate` describes `living body`, `rational` describes `rational animal`, and so on. This description is called *constitution* (συστατική, lat. constitutiva) [29, p. 179]. Porphyry thus makes a more precise distinction between nouns or extensional and adjectives or intensional concepts than Seneca and adds further relations to the logical connectives, which we will define more precisely in the following sections using the tree diagrams.

3 Typical Diagrams of S- and P-Trees

Tree diagrams have been handed down to us at least since the early Middle Ages. Although philosophers, theologians and scientists have dealt with these diagrammatic structures for many centuries, with the end of traditional Aristotelian logic in the modern age, more intensive occupations with the historic diagrams have become rare.

The studies by Barnes and Mansfeld are, in my view, the best to be found on our subject, even if I am not convinced about every detail. Barnes and Mansfeld argue that both text passages quoted in Sect. 2 are canonical for tree diagrams, but neither manuscript of these classical texts shows a tree diagram. Barnes and Mansfeld thus provide an ideal type, a kind of average or standard tree of the kind most often drawn between the early Middle Ages and the modern age to visualise the texts of Seneca or Porphyry. This can be easily seen by comparing the two trees with the historical illustrations in e.g. [46], [14], [40].

When any tree diagram was first drawn is a strong point of debate among historians of logic or art, but should not concern us here. For our main topic, it is first more crucial that Barnes and Mansfeld agree that there are two types of tree diagrams, one corresponding to the Senecaic text and one to the quotes of Porphyry. Both use similar logical connectives, differing in their structure. According to Barnes, the Porphyrian tree (P-tree) looks like Fig. 1a [29, p. 110] and according to Mansfeld, the Senecaic tree (S-tree) looks like Fig. 1b [25, p. 96].

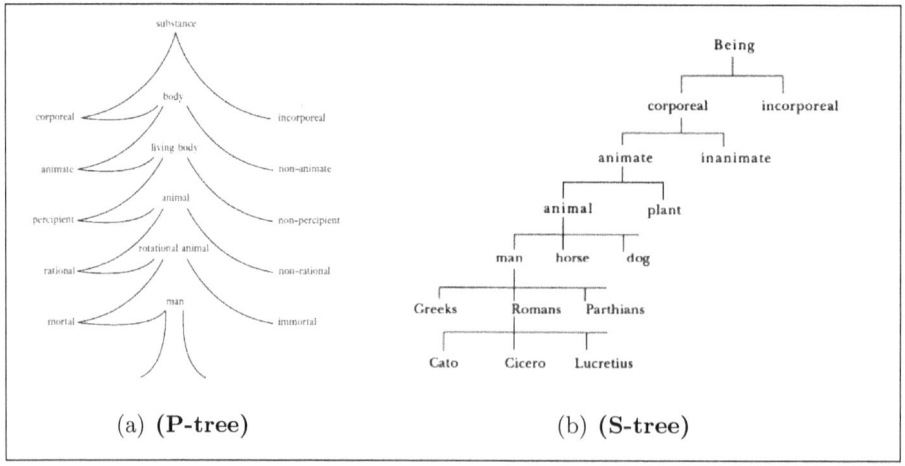

Figure 1: Tree Diagrams

Fig. 1a is made by Barnes, Fig. 1b by Mansfeld, and both figures correspond as far as possible to the diagrams that have been found again and again in many centuries since the Middle Ages. I therefore take both diagrams as the basis for the following analysis. In both diagrams, one has to take into account that Barnes and Mansfeld have considered more text than was discussed here in Section 2: Therefore, the S-tree of Fig. 1b shows more concepts than discussed in the quote of Sect. 2.1 and the P-tree of Fig. 1a also shows branches that do not correspond to the quotes discussed in Sect. 2.2. Note that the S-tree only shows the concepts that Seneca also discusses in his text, thus omitting many subconcepts that could have been developed on many genera on the right side of the diagram.

Nevertheless, the correspondence between the quotes from Sect. 2 given above and the respective trees in Fig. 1 should be apparent. For example, one can accurately identify in Fig. 1 the technical terms from Sect. 2: (1) In both diagrams, the *genus generalissimum* or *genus generale* is at the top, (2) followed by several levels of subspecies (*inter alia* or *genera specialia*). (3) The P-tree shows the concept **man** as *species speciallissima*. As stated above, however, Seneca does not introduce a technical term like *species specialissma* in his text. If he had introduced it, it would indicate the concepts **Greeks, Romans, Parthians** in the S-tree. (4) The P-tree does not include *individuals* that would have to be integrated at the trunk or roots of the tree.[1] (In almost all traditional textbooks including P-trees individuals such as **Socrates, Plato, Petrus** are given at this position.) The S-tree shows *singuli* at the bottom such as included in **Romans**, e.g. **Cato**, etc.

At the first glance, there is a clear diagrammatic difference between S-trees and P-trees: S-trees indicate the division of the genus generale into at least two extensional subspecies. One has to take into account that the first two levels in the S-trees under **being** are, strictly speaking, **corporeal being, incorporeal being** as well as **animate body, inamimate body**. Thus, on the one hand, all concepts mentioned in the S-tree are extensional. On the other hand, one would not have any problems showing the subtypes in the diagram for **incorporeal being**.

This is different in the P-trees. Starting from the top, the extensional concepts of the genus generalissimum and inter alia are divided

---

[1] Please also note that in Fig. 1 not only the individuals at the bottom of the diagram are missing, but that it should also read 'rational animal' and not 'rotational animal'.

into exactly two intensional subconcepts. In almost all the diagrams I know, there is a positive intensional side and a negative intensional side. In Fig. 1a, the positive intensional side is on the left and includes the concepts `corporeal, animate, percipent, rational, mortal`. The negative intensional side is on the right in Fig. 1a and shows the concepts `incorporeal, non-animate, non-percipent, non-rational, immortal`. Only the concepts in the middle of the diagram, i.e. between `substance` and `man`, are extensional. The concepts in the P-tree are thus mostly intensional and only partly extensional. As we will see in a moment, this gives rise to various difficulties in extending the P-tree.

4 Modern Interpretation of S- and P-trees

Our aim is not only to consider diagrams as good visual tools, but also to be able to give as exact a definition of them as possible, so that we can examine the differences between canonical diagrams more closely. In though we do not intend to develop a formal logical system with tree diagrams, it is still useful to follow Shin's method and distinguish between the syntax and the semantics of diagrams [38].

The *syntax* of a tree diagram can largely be described by using graph theory. Hacking [16] had already not only made a historical connection between the diagrams of ontology and graph theory, but also made some considerations about graph-theoretical interpretation, which, however, must be much more detailed. The *semantics* of a tree diagram is determined by the concepts whose relation can be visualised by a graph. In simplified terms, by syntax we mean here the form and appearance of the tree diagram, by semantics the meaning of all parts of the diagram. In sum, a tree diagram is a representation of concepts and their relations with the help of a graph.

So far, we have mainly worked with trees whose syntax was determined by the semantics of Seneca's and Porphyry's text. Barnes and Mansfeld provided two typical visualisations of the two texts, which were semantically occupied with the concept used by Seneca or Porphyry. In the following, however, let us try to look at the graph of the diagrams.

4.1 Syntax of Trees

We begin by presenting a set of simplified definitions from graph theory, taking [11] as our guide:

Graph. Let G = (V,E) be a *graph* if V is a finite set of vertices or nodes and E is a set of relations on V represented as edges. Two graphs G_1 and G_2 are called *isomorphic* if they are structurally the same. *End vertices* in a graph G are the two vertices x and y if they are connected by an edge xy or yx. Two edges are *connected* if they have a common vertex, and two vertices are connected if they have a common edge. The *degree* deg(V) denotes the number of vertices connected to a vertex. The *input degree* deg-(V) of a vertex V is the set of edges leading to this vertex. The *output degree* deg+(V) of a vertex V is the set of edges leading away from this vertex. In an *undirected* graph, the edges xy and yx are equal, $\overline{xy}$ for short. In a *directed* graph, the edges xy and yx are unequal, $\overleftarrow{xy}$, $\overrightarrow{xy}$ for short. In an *edge-weighted graph*, each edge is assigned a real number. G' is a *subgraph* of G if G' is a graph and every set of G' is a real subset of G, i.e. $V(G') \subseteq V(G)$, $E(G') \subseteq E(G)$. In this case, G is also called a *supergraph* of G'. A *path* P is a composite of vertices whose *length* k is denoted by the number of connected vertices, i.e. P^k. A path with connected vertices a, b, c, d would then be $P =$ abcd. A path P that contains two times a vertex connected by at least two edges is called a *circle*. If a, b, c are vertices, $P =$ abca is a circle. A *line* L is a path over several vertices and edges, where each vertex is connected to another by only one edge and the direction of each edge is continued by the next one.

Tree. A graph G is called a *tree* T if it is connected and contains no circles. In T there can exactly be one *root* that is a node of degree 2 and does not form a line with the end vertices of its edges. Any tree with a root is a *rooted graph*. In T, a vertex that has a degree of exactly 1 is a *leaf*. If a vertex in T has a degree of 2 and forms a line with the end vertices of its edges, this vertex is called a *non-branching node*. In T, a vertex that has a degree of ≤ 3 is a *inner vertex*. In a directed graph, which is a tree, the vertex x in the edge $\overrightarrow{xy}$, is called a *child* and the vertex y is called a *parent*. An *out-tree* is a rooted directed graph where the root has an output degree of ≤ 2 and an input degree of 0, the leaves have an input degree of 1 and an output degree of 0 and where there is only one direct directed path to each leaf starting from a root. An *in-tree* is a rooted directed graph where the leaves have an initial degree of 1 and an input degree of 0, the root has an input degree of ≤ 2 and an output degree of 0, and where there is only one direct directed path from each leaf, ending at the root. If in an out-tree each inner vertex has the same output n and an input of 1 or in an in-tree each inner vertex has an input of n and an output of 1, then it is called *regular*, otherwise irregular. A regular tree with n= 2 is called a *binary tree*, with n= 3 a *ternary tree*, with n= 4 a *quaternary tree*, etc. If the length of the path between each leaf and the root in a tree is always the same, the tree is called *balanced*. If the length of P is different, it is called *non-balanced*. A set of disjoint trees is a *forest*.

In contrast to e.g. formal logic, we find in traditional tree diagrams no rules of construction: We can create innumerable graphs, which we may also name as trees with the help of modern graph theory, but a traditional S- or P-tree has a special form, which roughly corresponds to either Fig. 1a or Fig. 1b. If we look at these ideal-typical trees of Fig. 1, we quickly realise that there is a syntactic iteration in both P- and S-trees. We therefore define these iterations as schemes as follows:

Schemes. P-trees and S-trees are two supergraphs composed of several isomorphic subgraphs, which are called *schemes*. A *P-scheme* consists of four vertices and three edges, an *S-scheme* of three vertices and two edges. In a P- and in a binary S-scheme there is one vertex (V_1) at the top that is named *top root* and two vertices (V_2, V_3), each of which is below V_1, so that V_2 is below V_1 on the left and V_3 is below V_1 on the right. In P- and S-schemes, $\overline{V_1 V_2}$ and $\overline{V_1 V_3}$ holds. In a P-scheme, there is a vertex V_4 that is vertically below V_1 so that it is approximately on the same plane as V_2 and V_3. For P-schemes holds $\overline{V_2 V_4}$.

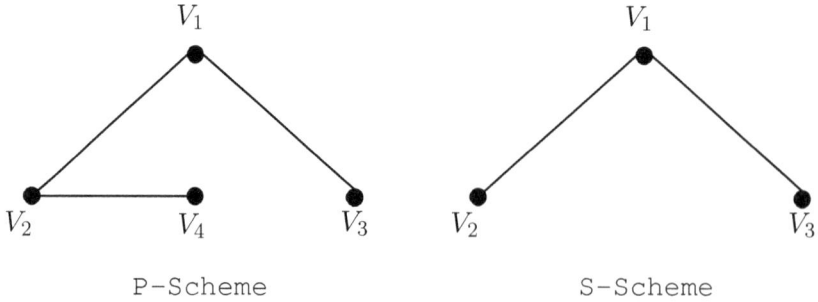

P-Scheme   S-Scheme

If S- and P-trees are nothing but repetitions of the same scheme (which only deviates from it at its upper and lower ends, as in Fig. 1), then we can give something like rules to construct iterations of these schemes.

Iteration. An *iteration* is a substitution of a schema at a certain vertex of an already existing schema. The graph of a P- or S-trees, which consists of at least two schemes, is named *P-iteration* or *S-iteration*. In P-schemes, only V_4 of an existing graph may be substituted by V_1 of a graph to be substituted if an outtree is to be constructed. In an intree, V_1 of an existing graph may be substituted by V_4 in a P-scheme. In S-schemes, only V_2 or V_3 of an existing graph may be substituted by V_1 of a graph to be substituted if an outtree is to be constructed. In an intree, V_1 may be substituted by either V_2 or V_3 in an S-scheme. The outtree thus grows upwards, the intree downwards.

Perhaps we have now already found a sufficient way to describe a large part of the trees of Fig. 1 with the existing syntax. Of course, it must be noted that we have only constructed binary S-schemes so far. In this respect, it must be added, for example, that each n-ary S-scheme increases the number of vertices and edges by 1 for n > 2. In this case, $V_1 V_n$ applies in each case.

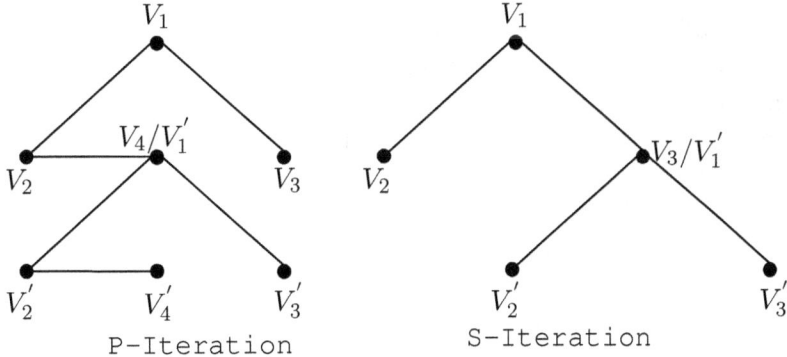

P-Iteration   S-Iteration

Furthermore, logicians have been pointing out since the Middle Ages that there are semantic relations between vertices that are often not even drawn in the traditional tree diagram (e.g. in [27]). In the case of P-trees, this concerns possible edges between two vertices in two schemes, in the case of S-trees possible edges between two vertices in already one scheme. To make this implicit information explicit, we now extend

S-schemes and P-iteration.

Extended iterations and schemes. Isomorphic subgraphs of a tree are called *extended* if they contain edges between vertices that are not defined by a scheme or an iteration. In this case, we speak of S-extensions or P-extensions. In an S-scheme it is possible to draw the edge $\overline{V_2 V_3}$. In a P-scheme it is possible to draw the edge between the vertices $\overline{V_1/V_4}$ and the vertices $\overline{V_3 V_4}$. In a P-iteration, edges can be drawn between all V_2 or all V_3 vertices, i.e. $\overline{V_2 V_2'}$ or $\overline{V_3 V_3'}$.

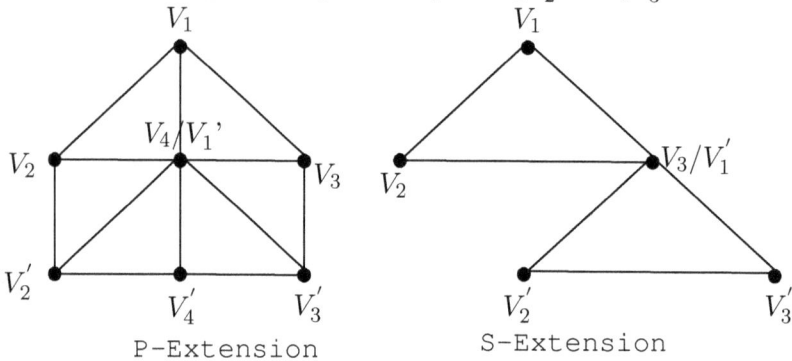

4.2 Semantics of Trees

The separation of syntax and semantics in tree diagrams is not quite simple. We have so far used graph theory as syntax. By semantics we can now understand the meaning of vertices and edges in a tree diagram: the meaning of a vertex is a concept, the edges between two terms is a relation. As Frické has argued, the relations within the tree are logical in nature, even if modern logic often dispenses with them [15].

We have already sent some important information about the concepts in Section 3, but we can now make them a little more precise. We will use Porphyry's classification of concepts from Table 2.

Concepts. In an S- or P-tree, the V_1 vertex that functions as the top root is the *genus generalissimum*. In a P-tree, any vertex that functions simultaneously as a V_4 in one schema and as a V_1 in another schema is an *inter alia* concept. Any vertex that is V_4 but is not substituted again by another schema and thus does not function again as V_1 in another schema is a *species specialissima*. In P-schemes, the vertex under the species specialissima are the *individua*; but there is no uniform syntactic notation for drawing individuals. In S-schemes, any vertex that functions simultaneously as V_2 or V_3 in one schema and as V_1 in another schema is an *inter alia* term. Any vertex that is V_2 or V_3 but is not substituted again by another schema and thus does not function again as V_1 in another schema is an *individual*. In an S-tree, any inter alia vertex that is directly connected by edges to vertices of the individuum is a *species specialissima*.

We can add that a genus generalissimum in P- and S-trees is usually called **substance** or **being** or a synonym of these concepts. However, as described in the introduction, recent research has shown that there are indeed more semantic forms. In any case, all V_2 terms in a P-tree are adjectives with positive connotations, all V_3 terms are adjectives with negative connotations (*intensional concept*). All V_4 concepts in a P-tree are positive nouns (*extensional concept*). S-trees usually contain only one grammatical form of concepts, even if Seneca mixes adjectives and nouns, as described above.

There is usually a semantic relation between most of the concepts in a tree. We have already taken these relations in Sect. 2 from the texts of Seneca and Porphyry, namely: divsion, subordination, exclusive disjunction, constitution.

Relation. The edge relation between $\overline{V_1 V_2}$ and $\overline{V_1 V_3}$ in a P- or S-scheme of an outtree is the *division*. If it is an intree, both relations are *constitutions*. The edge relation $\overline{V_3 V_4}$ in a P-scheme and $\overline{V_2 V_3}$ in a S-scheme is an *exclusive disjunction*. In P-schemes, the edge relation $\overline{V_2 V_4}$ is a *constitution*. The edge relation $\overline{V_1 V_4}$ in a P-scheme as well as V_2/V_2' and $\overline{V_3 V_3'}$ in a P-iteration is a *subordination*. *Division* is the separation of either a genus into at least two species or of a species into at least two individuals. *Constitution* is the semantic correspondence between an adjective and a noun. *Exclusive disjunction* denotes the fact that if the concept of one vertex is true, the other is false (taking into account only the relationship to each other and not to other concepts). *Subordination* denotes the relation of a genus to its next lower species, provided that both concepts are either extensional or intensional.

It is already evident from the above-mentioned relations that there is a certain order, which we determine more precisely below under Suppes's definition ([43, p. 210ff.]). This also allows us to create an ordered graph

77

of the traditional trees:

Ordered relations. *Subordination* and *division* are transitive, irreflexive and not symmetrical: If $\overline{V_1V_4}$ are in subordination and so are $\overline{V_4V_4'}$, them $\overline{V_1V_4'}$. If $V1$ is genus for a kind V_x, then $V_x = V_1$ cannot hold. For the *subordinations* in a P-scheme or a P-iteration this means $\overrightarrow{V_1V_4}$, $\overrightarrow{V_2V_2'}$, $\overrightarrow{V_3V_3'}$. If the *division* $\overline{V_1V_2}$ and $\overline{V_1V_3}$ applies, $\overline{V_1V_2'}$ must also apply if also the division $\overline{V_3V_2'}$ and $\overline{V_3V_3'}$ applies. If V_1 divides into V_x and V_y, then it must hold that $V_x \neq V_1$ and $V_y \neq V_1$. If V_2 and V_3 are divided by V_1, then V_1 cannot be divided by either V_2 or V_3, so $\overrightarrow{V_1V_2}$ and $\overrightarrow{V_1V_3}$ hold. Under the same condition, however, the *constitution* relation $\overrightarrow{V_2V_1}$ and $\overrightarrow{V_3V_1}$ holds. Since nothing contradicts the self-constitution of a concept, constitution is transitive and reflexive. The ordered relation of the *exclusive disjunction* is known (symmetrical, irreflexive), but it must be pointed out that there is a certain transitivity in P-schemes: If the exclusive disjunction $\overline{V_3V_4}$ and the *constitution* $\overline{V_2V_4}$ exist, then the exclusive disjunction $\overline{V_2V_3}$ also applies. Something similar also applies in S-schemes if they have at least 3 schemes and are balanced: If the *divisions* $\overline{V_2V_2'}$, $\overline{V_2V_3'}$ as well as $\overline{V_3V_2''}$, $\overline{V_3V_2''}$ apply, then the *exclusive disjunctions* $\overline{V_2'V_3'}$ and $\overline{V_2''V_3''}$ must also apply. If this is the case, $\overline{V_3'V_2''}$ also applies, which can be expressed syntactically by a further edge, but implicitly also the exclusive disjunction of all further vertices among each other on the same level.

5 Problems and Examples

We have now made an offer in Section 4 to be able to analyse tree diagrams better. However, this is not without problems and, moreover, concepts without intuition remain, as Kant says, empty. We will therefore first discuss some problems in this section and then go into some examples to test the definitions and results of the previous sections. This allows us to fill the empty definitions of the previous chapters with intuitions.

Since there are clear differences between traditional trees and the terms of graph theory, in the following we will only explicitly refer to them when traditional trees and not trees in the sense of graph theory are meant.

5.1 Problems with Trees

There are some similarities, but also many differences between today's graph theory and the traditional trees in the vein of Seneca and Porphyry. Only a few comparisons and observations can be made here before we sketch some examples:

(1) An icon resembles an object, without having to be physical itself, to a certain degree [42]. Symbols need not bear any resemblance to what they represent. By using such definitions, we can yet say that traditional trees are iconic, whereas trees of graph theory are usually symbolic, or iconic only by accident. This can be seen, for example, in the fact that in the traditional trees the root is at the bottom of the diagram, the leaves always at the side and the tree top always at the top of the diagram. In graph theory, the root of a tree is usually at the top of the figure, while the leaves are at the bottom. Nevertheless, trees in graph theory can have any conceivable shape, provided they are circle-free.

(2) The method of perductio described in Seneca corresponds to an in-tree in graph theory, the method of dihairesis corresponds to an out-tree. But not every intree or outtree of graph theory corresponds to a traditional tree.

(3) The P-trees and S-trees shown in Fig. 1 are not isomorphic. Both are unbalanced trees because the length of the paths between root and leaves is shorter on the left side than on the right side, but the structure of both trees is different. If one were to arrange the P-tree vertically like an S-tree, it would be more noticeable that the adjectives or intensional expressions form a half-leaf that is missing in the S-tree.

(4) As mentioned above, several relations are missing in the illustrations of Fig. 1, which are described in Seneca and Porphyry and can also be found in some illustrations of P- and S-trees from the Middle Ages onwards. If these relations were added, from a graph-theoretical point of view, the P- and S-trees would no longer be trees at all, because the graphs would have circles. We will see this at the end of this section in the extended graphs.

5.2 Examples of P-Trees

In this subsection, we will take two P-trees as examples, analyse them using the syntax and semantics mentioned above, and include some interpretations of classical texts. The two examples are P-T1 and P-T2.

In the following, we will weight the edges in order to abbreviate the relations: division (1), subordination (2), exclusive disjunction (3) and constitution (4). We will not fill the vertices conceptually, as this can be done, for example, with the help of the comparison with Fig. 1.

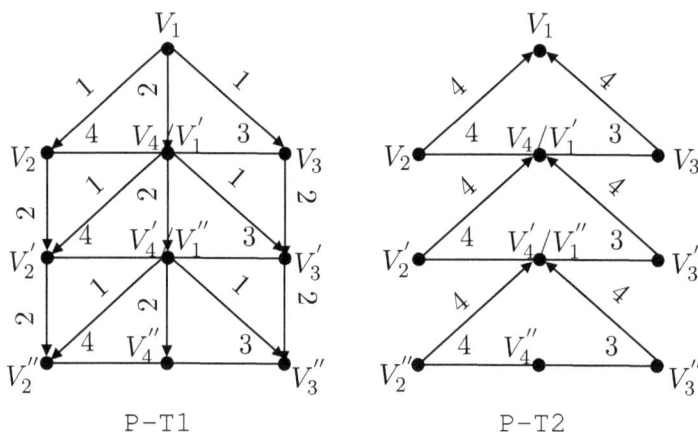

P-T1   P-T2

P-T1 and P-T2 show graphs that can be found in a similar way in several textbooks of the early modern period, e.g. in [13, p. 84] or [35]. We see three subordination paths in P-T1, which are often regarded as the most important information, as they visualise the relation between homogeneous concepts (extensional or intensional, positive or negative concepts). If we take out the relations between heterogeneous concepts, i.e. the edge weights 1, 3, 4, we are left with a forest of three trees, which is displayed in P-F1. We have in F1 the disjoint union of the trees P-T1$'$, P-T1$''$, P-T1$'''$, where P-T1$'$: $\{V_2, V_2', V_2''\}$ indicates the positive intensional, P-T1$''$: $\{V_1, V_1'/, V_1''/V_4', V_4''\}$ the positive extensional and P-T1$'''$: $\{V_3, V_3', V_3''\}$ the intensional negative concepts. P-T1$''$ is traditionally referred to as *linea directa* and P-T1$'$ and P-T1$''$ as *linea indirecta* [7, p. 276, p. 400].

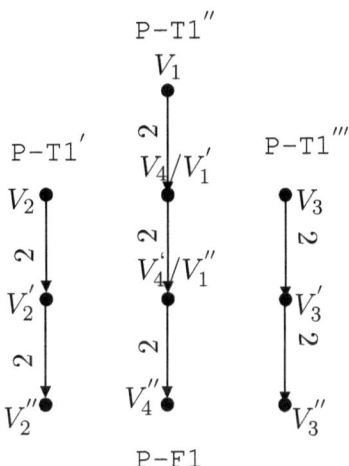

P-F1

However, if one focuses in P-T1 on the relations, a process is often highlighted in the outtree P-T1, which runs top-down through a directed tree from V_1 to V_4. With this focus, the negative side of P1 is completely ignored and the constitutional process is interpreted as a directed vertex in the P-scheme $\overrightarrow{V_2V_4}$. The result of this process focus of the outtrees is the subgraph P-Sub1, where V_1 is called *terminus a quo* and V_4'' *terminus ad quem*. Traditional concepts of the description of the top-down process, going from the genus generalissimum to the individua or species specialissima, are κάθοδος, παραγωγή, deductio, and many others.

The exact opposite concept is called ἄνοδος, ἐπαγωγή, inductio, etc., and is visualised by the directed graph P-Sub2, which is a subgraph of P-T2, where the constitution relation $\overrightarrow{V_4/V_1V_2}$ has been directed.[2] In P-Sub2, V_4 is now the *terminus a quo* and V_1 the *terminus ad quem*. So the process is *bottom-up*, from individua or species specialissma to genus generalissimum. The two subgraphs and the description of the associated processes can be found, for example, in [47, p. 309].

We see in both tradional tree graphs that they have an iconicity that many modern graphs that are trees do not have: As Barnes says, the P-trees remind us above all of pine trees [29, p. 110]. In P-T1, for example, one can clearly see the trunk in the middle, i.e. P-T1″ in P-T2 one can see mainly the fir shape of the branches. These terms, however,

---

[2] A history of concepts and ideas of these bottom-up and top-down processes from antiquity to modern philosophy of science can be found in [23].

do not correspond to the definitions of modern graphs any more than the term 'forest' does to traditional trees.

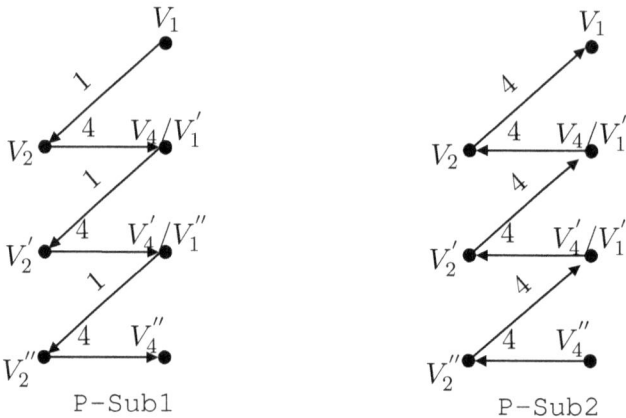

P-Sub1   P-Sub2

5.3 Examples of S-Trees

In this subsection, I will take an S-tree S-T1 as an example, analyse it using the syntax and semantics mentioned above, and include some interpretations of classical texts.

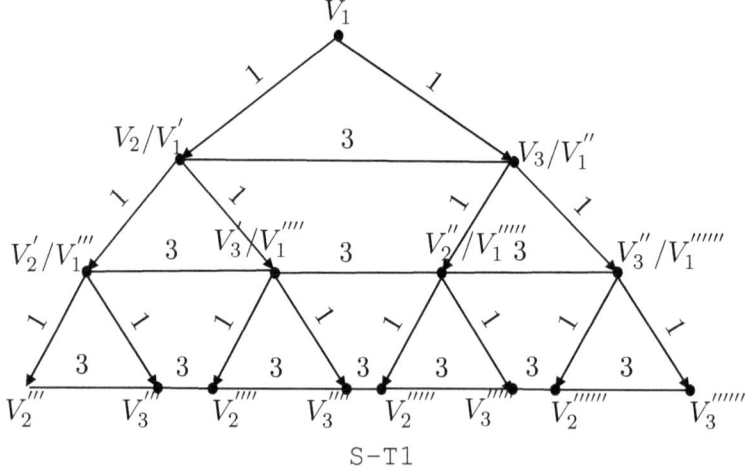

S-T1

Like P-T1 and P-T2, S-T1 is not a tree in the sense of graph theory. However, if one were to remove the edges for the exclusive disjunction (3), S-T1 would be a balanced outtree. If the directionality of the edges were then reversed, the outtree would become an intree in which the edge weight 1 no longer stands for the exclusive disjunction, but for the constitution. This can be seen in some interpretations of the early modern period, in which constitution was represented by the inclusive disjunction: *aut...aut*, then becomes *vel...vel* in edge-weighting 3.[3] In the latter case, the moment of constitution of edge-weighting 1 rests on the fact that V_2 and V_3 can both hold in V_1. Thus, the interpretation of the disjunction relation also has an impact on the constitution relation in bottom-up processes.

In any case, V_1 represents the genus generalissimum in S-T1, the vertices V_2V_1' and V_3V_1'' being the inter alia, $V_2'V_1''' \ldots V_3''V_1''''$ the species specialissima and $V_2''' \ldots V_3'''''$ the individua. The relations are limited to the dihairesis (1) and the exclusive disjunction (3). In particular, through the dihaireses, the process is seen as top-down, with the genus generalissimum being the terminus a quo and the individua the terminus ad quem.

6 Conclusion and Outlook

In this paper, it was shown what the traditional trees in philosophy looked like, which were used especially in the roughly mentioned period between the 9th century and the 19th century. First, we analysed the relevant texts of Seneca and Porphyry and, using the ideal-typical interpretation of Barnes and Mansfeld. Then, we showed how these ancient texts were diagrammatically reworked. In the process, several difficulties were identified that have so far stood in the way of a modern interpretation of these trees.

Nevertheless, a method was proposed in Section 4 to investigate the syntax and semantics of traditional tree diagrams. It has been shown that graph theory offers a possibility for interpreting the syntax, whereas for the semantics one has to explain above all what the meaning of the vertices and edges is, i.e. concepts and (partially) logical relations. It must be emphasised that traditional trees are organized by the same structures that are repeated again and again: We have therefore spoken

---

[3]Compare, for example, the relation of 3 in P. Ramus, who sees it as an exclusive disjunction [32, pp. 95ff.], with the relation of 3 in the S-tree of B. Keckermann's, who interprets it as an inclusive disjunction [20, l. I, pp. 5f.].

of schemata and of iterations. In addition, however, graph theory also offers the possibility of representing the otherwise only implicit information of relations in the form of edges.

The greatest difficulties in the modern interpretation arise from the fact that the concept of tree is used differently in traditional ontology than in modern graph theory. Although modern graph theory also grew out of metaphysics, the modern definition of the concept of tree no longer emphasises iconicity: Modern trees of graph theory no longer need to have the form of a tree, although there are also problems with traditional iconic diagrams, for example, that in outtrees growth is from the tip to the root.

The modern interpretation proposed here, however, has attempted to bring the traditional tree diagrams closer to graph theory again, that is, a use of graph-theoretical definitions in the language of analysis. To what extent each detail proposal must be considered a success remains to be seen. However, the examples shown in Section 5 should be sufficient evidence to show that the definitions developed in Section 4 provide an effective means of constructing and analysing traditional trees.

From a graph-theoretical point of view, the success of this paper is likely to be very manageable. From a philosophical point of view, the methods proposed here could offer a way to present certain historical and systematic topics in a new light. As examples, consider some topics related to traditional tree diagrams. For example, philosophers such as Murmellius and Sfondrati made syntactic and semantic extensions to diagrams such as P-T1 in order to carry out holistic language analysis [5]; Johann Christian Lange designed the first logic machine on the basis of tree diagrams [24, chap. 2.2.3]; in Kant's or Hegel's systems tree diagrams play implicitly an essential role [31], [45, §6]; Tree diagrams had an influence on modern logics, as can be seen in Peirce and Gentzen for example [1]; and today's ontologies in the field of knowledge representation tie in with the methods of traditional tree diagrams [41].

References

[1] I. H. Anellis, F. F. Abeles, The Historical Sources of Tree Graphs and the Tree Method in the Work of Peirce and Gentzen, in *Modern Logic 1850 –1950, East and West*, Springer, Cham, 2016, pp. 35–97.

[2] M. Armisen-Marchetti, 'Ontology and Epistemology', in A. Heil, G. Damschen (ed.) *Brill's Companion to Seneca*. Brill, Leiden, 2014, pp. 217–238.

[3] S. J. Russell, P. Norvig, *Artificial Intelligence. A Modern Approach*. 3rd ed. Pearson, Boston, 2016.

[4] J.-Y. Béziau, 'The New Rising of the Square of Opposition', in J.-Y. Beziau, D. Jacquette (eds.) *Around and Beyond the Square of Opposition*. Birkhäuser, Bâle, Suisse, 2012, pp. 3–19.

[5] P. R. Blum, *Studies on Early Modern Aristotelianism*. Brill, Leiden 2012.

[6] C. Chakraborti, *Logic: Informal, Symbolic and Inductive*. Prentice Hall, New Delhi, 2007.

[7] J. Channevelle, *Accurata totius philosophiae institutio juxta principia Aristotelis*. Martin, Paris, 1667.

[8] L. Demey, 'The Porphyrian Tree and Multiple Inheritance. A Rejoinder to Tylman on Computer Science and Philosophy' in *Foundations of Science*, 2018, 23, pp. 173–180.

[9] L. Demey, 'From Euler Diagrams in Schopenhauer to Aristotelian Diagrams in Logical Geometry' in J. Lemanski (ed.) *Language, Logic, and Mathematics in Schopenhaue*, Birkhäuser, Cham, 2020, pp. 181–205.

[10] K. Desmond, *Music and the moderni, 1300–1350: The ars nova in Theory and Practice* Cambridge, Cambridge Univ. Press, 2018.

[11] R. Diestel, *Graph Theory*, Springer, Heidelberg, New York, 2005.

[12] G. Englebretsen, *Robust Reality: An Essay in Formal Ontology*. ontos, Frankfurt et al., 2012.

[13] Eustachius a Sancto-Paulo, *Summa philosophiae quadripartita, de rebus dialecticis, tom. 1*, Chastellain, Paris, 1609.

[14] A. Even-Ezra, *Lines of Thought Branching Diagrams and the Medieval Mind*. Chicago Univ. Press, Chicago, 2021.

[15] M. Frické, 'Logical Division', *Knowledge Organization* 43:7 (2016), pp. 539–549.

[16] I. Hacking, 'Trees of Logic, Trees of Porphyry' in J. Heilbron (ed.) *Advancements of Learning: Essays in Honour of Paulo Rossi*. Olschki, Firenze, 2006, pp. 221–263.

[17] L. Geldsetzer, *Logical Thinking in the Pyramidal Schema of Concepts: The Logical and Mathematical Elements*. Springer, Dodrecht, 2013.

[18] J. A. Groenland, 'Humanism in the Classroom, a Reassessment', in R. Bod, J. Maat, T. Weststeijn (ed.), *The Making of the Humanities: Volume 1: Early Modern Europe*. Amsterdam Univ. Press, Amsterdam, 2010, pp. 199–230.

[19] L. Jansen, 'Classifications', in K. Munn, B. Smith (ed.) *Applied Ontology: An Introduction.* ontos, Heusenstamm, 2008, pp. 159–173.

[20] B. Keckermann, *Systema logicæ.* Stöckle, Francofurti, 1628.

[21] J. C. Lange, C. Weise, *Nvclevs Logicae Weisianae: [...] illustrates [...] per varias schematicas [...] ad ocularem evidentiam deducta [...]*, Müller, Giessen (Gissae-Hassorum), 1712.

[22] J. C. Lange, *Inventvm novvm quadrati logici vniversalis*, Müller, Giessen (Gissae-Hassorum), 1714.

[23] J. Lemanski, *Summa und System*, mentis, Münster, 2013.

[24] J. Lemanski, *World and Logic*, College Publications, London, 2021.

[25] J. Mansfeld, *Heresiography in Context. Hippolytus' Elenchos as a Source for Greek Philosophy.* Brill, Leiden, 1992.

[26] A. Moktefi, S.-J. Shin, 'A History of Logic Diagrams', in D.M. Gabbay, F.J. Pelletier, J. Woods (eds.) *Logic: A History of its Central Concepts.* Elsevier, Burlington, 2012, pp. 611–683.

[27] J. Murmellius, *In Aristotelis decem praedicamenta isagoge* Hieronimus Calepinus, Venetiis, 1513.

[28] A. Newell, C. Shaw, H. Simon, 'Empirical Explorations of the Logic Theory Machine: A Case Study in Heuristics', in *RAND Corp. Report P-951* (March 1957).

[29] Porphyry, *Introduction. Translated, with a Commentary, by Jonathan Barnes.* Clarendon, Oxford, 2003.

[30] H. Prade, P. Marquis, O. Papini, 'Elements for a History of Artificial Intelligence, in Ibid. (ed.) *Guided Tour of Artificial Intelligence Research. Vol. 1: Knowledge Representation, Reasoning and Learning.* Springer, Heidelberg, 2019.

[31] V. Pluder, J. Lemanski, 'A Diagrammatic Representation of Hegel's 'Science of Logic", in G. Stapleton, A. Basu et al. (eds.), *Diagrams 2021: Diagrammatic Representation and Inference.* Springer, Cham, pp. 255–259.

[32] P. Ramus, *Professio regia, hoc est septem artes liberale.* Henricpetri, Basel, 1576

[33] D. Reichling, *Johannes Murmellius: sein Leben und seine Werke.* Herder, Freiburg, 1880.

[34] W. Risse, *Die Logik der Neuzeit, Vol. 1: 1500–1640*, frommann-holzboog, Stuttgart-Bad Cannstatt, 1964.

[35] Philipp a Sanctis, *Summa philosophica ex mira principis philosophorum Aristotelis*, Friess, Cologne, 1665.

[36] F. Schang, 'Oppositions and opposites', in J.-Y. Beziau, D. Jacquette

(eds.) *Around and Beyond the Square of Opposition*. Birkhäuser, Bâle, Suisse, 2012, pp. 147–173.

[37] Seneca, *Ad Lucilium Epistulae Morales. With an English Translation by Richard M. Gummere*. Vol. 1. Harvard Uni. Press, Cambridge, Mass. 1917 (Repr. 1953).

[38] S.-J. Shin, *The Logical Status of Diagrams*. Cambridge University Press, Cambridge 1995.

[39] C. Sfondrati, *Logica*. Jacob Müller, Monasterium S. Gallen, 1696.

[40] S. Siegel, *Tabula. Figuren der Ordnung um 1600*. Akademie, Berlin 2009.

[41] J. S. Sowa, *Knowledge Representation. Logical, Philosophical, and Computational Foundations*. Brooks/Cole, Pacific Grove, CA, 2002.

[42] F. Stjernfelt, Diagrams as a Centerpiece of a Peircean epistemology, in *Transactions of the Charles S. Peirce Society* 36(3): pp. 357–384.

[43] P. Suppes, *Introduction to Logic*. Van Nostrand Reinhold, New York et al. 1957.

[44] A. Schumann, *Behaviourism in Studying Swarms: Logical Models of Sensing and Motoring*. Springer Nature, Cham et al. 2019.

[45] C. Tolley, 'The Generality of Kant's Transcendental Logic', in *Journal of the History of Philosophy* 50:3 (2012), pp. 417–446.

[46] A. R. Verboon, *Lines of Thought. Diagrammatic Representation and the Scientific Texts of the Arts Faculty, 1200–1500*. PhD thesis, Leiden, 2010. Weger

[47] L. Weger, *Mentis Prima Operatio Logica* Segebadius, Regiomontium, 1630.

[48] J. Wildberger, *Seneca und die Stoa: Der Platz des Menschen in der Welt*. Vol. 2. de Gruyter, Berlin, 2016.

THE NUMBER OF CATEGORICAL PROPOSITIONS IN BOETHIUS'S COMMENTARIES ON ARISTOTLE'S *PERI HERMENEIAS*

MANUEL CORREIA
Pontifical Catholic University of Chile
mcorreia@uc.cl

Abstract

This paper argues that Boethius and Ammonius, the most important commentators of Aristotle's logic during the VI AD, are moved by Aristotle's *Peri Hermeneias* to identify and classify every categorical or simple proposition and its contradictory opposite. The exegesis, which includes an arithmetical count of these categorical propositions and their opposite pairs, aims at giving completeness to Aristotle's theory of negation, and it is based on eight logical properties, namely, quality, quantity, modality, matter, indefinite subject, indefinite predicate, singular terms, and tense. The aim of the paper is to explain the value of this exegesis and to suggest that the history of Western logic has been the history of the study of these eight properties.

1 Introduction

There are some passages where Aristotle propounds on calculating the number of elements contained in a genus. For example, *Politics* 4, 1290b21 ff., *De Gen* 2, 330a30, *Met.* 4, 1014a5, and *Parts of Animals* 643a16-24.[1] In *Politics*, by making an analogy with the parts of animals,

---

[1] In [3, *Met.* 1014a5 ff.], Aristotle tries to get the number of varieties (*trópoi*) involved in the notion of cause (*aitía*). In [3, *Parts of the Animals* 643a and ff.],

he asks for calculating the number of all possible forms of government once accepted that every state consists of many parts.² Also, it is very clear his intention in [3, *De Gen*], when he calculates the number of all possible pairs contained in the set of four basic natural qualities of tangible beings (humid, dry, hot, cold). Here he first asserts correctly that the arithmetical result of combining 4 elements is 6. However, since he disregards pairs with contrary qualities (such as dry and humid, and cold and hot), he correctly states that there are only 4 consistent combinatory pairs.³

These passages seem to suggest an intellectual disposition, which is confirmed by another passage, if the ancient commentaries on Aristo-

---

Aristotle rejects the dichotomous division as a research instrument to obtain the *infima* or indivisible species of a genus. He advances the criticism that it would be necessary to accept the absurd consequence that, if one follows this method, one should accept that the number of species of animal (or any other species of a genus) is always even. The passage shows that Aristotle was familiar with the idea of counting the species of a genus. "It is plain then that we cannot get at the indivisible species of the animal, or any other kingdom, by bifurcate division. If we could, the number of ultimate differentiae would equal the number of indivisible animal species. For assume an order of beings whose prime differentiae are White and Non-white. Each of these branches will bifurcate, and their branches again, and so on till we reach the differentiae, whose number will be four or some other power of two and will also be the number of the ultimate species." (Trans. W. Ogle).

²The calculus, Aristotle adds, can be attempted as if we were going to determine the different species of animals starting from their indispensable parts: "I have said that there are many forms of government, and have explained to what causes the variety is due. Why there are more than those already mentioned, and what they are, and whence they arise, I will now proceed to consider, starting from the principle already admitted, which is that every state consists, not of one, but of many parts. If we were going to speak of the different species of animals, we should first determine the organs which are indispensable to every animal, as for example (...)". See [3, *Politics* IV, 4, 1290b. 21-27; trans. B. Jowett].

³In [3, *De Gen* II, 3, 330a30-330b1], Aristotle says: "The elements are four, and any four terms can be combined in six couples." After that, as each element to be paired holds its natural quality (so water is moist, earth dry, etc.), Aristotle reduces the combinations from 6 to 4, because contraries cannot be paired: "Contraries, however, refuse to be coupled; for it is impossible for the same thing to be hot and cold, or moist and dry. Hence it is evident that the couplings of the elements will be four: hot with dry and moist with hot, and again cold with dry and cold with moist." (trans. H. H. Joachim).

tle's *Peri Hermeneias* are right.[4] According to them, Aristotle's *PeriH* would attempt to identify and classify every species of categorical proposition along with counting their number to be certain that all the propositional forms have been distinguished.

Authors from II AD, namely, Alexander of Aphrodisias and Galen,[5] confirm that Aristotle's logic is familiar with this theoretical issue. But it is not until the Neoplatonic school, especially the period between IV-VI AD, which is well represented by the commentaries of Boethius [6] and Ammonius Hermeias[7] that this problem becomes central in the teaching of Aristotelian logic.

However, medieval commentators did not focus on the issue. Byzantine school tradition has a later writing (*Anonymous Heiberg*) where the calculus of propositions appears again. His anonymous author subdues this count to find the number of syllogisms, which seems to be correct. However, although interesting, Barnes [2, p. 136], is right when concluded that this calculus is plagued with many errors and an obscure methodology.

In modern times, the young Leibniz, in his [8, 13, II], refers to Aristotle's passage of [3, *De Gen* 2, 330a30] to demonstrate that the search for an arithmetical calculus of the elements of a set had an important precedent in Aristotle. He is not aware of the exegesis of the ancient commentators of Aristotle [7, pp. 232-240], but he is, according to my limited view, the last great thinker in relating this theoretical issue to Aristotle.

Contemporary interpreters, when paying attention to this propositional calculus, have rather dismissed its importance under an admitted

---

[4] Along with the article, Aristotle's *Peri Hermeneias* will be shortened by *PeriH*. Cf. [13].

[5] In [15], Alexander of Aphrodisias strives to determine the number of moods of the categorical syllogisms in each figure. For the First Figure, in which the point is clearer, cf. [1, pp. 113-114]. We also know that Galen in [10, xvii, p. 51] refers to his (now lost) book *On the Number of Syllogisms*, in which the calculus of this number would be (presumably) clarified. Cf. also [11].

[6] Boethius reports Syrianus's calculus. Cf. [12, 2, 20-24, pp. 321-324]

[7] Ammonius follows the oral teaching of Proclus. Cf. [5, p. 90, 21-p. 91, 3; p. 160, 17-32; p. 218, 30-p. 219, 24].

idleness and even absurdity,[8] and they have contributed to increasing the belief that ancient schools contribute superficially to clarify Aristotle's intention.

Ancient and contemporary authors are in clear opposition. This advises us to come back to the ancient commentaries written in the VI AD, those by Boethius, Ammonius, and Stephanus,[9] and to search what could be the reason why they unanimously accept the importance of identifying, classifying, and counting the species of proposition. The aim of this paper, then, is to explain why this ancient exegesis was important for Aristotelian logic and how it can agree with contemporary interpreters in the concern of a sound and complete theory of logic. Further, the paper will suggest that this exegesis is still useful as far as the history of logic has been a study of the properties allowing the classification and count of simple or categorical propositions.

2 The two-term propositions

According to Boethius's commentary, after dealing with truth and falsity in contingent propositions about the future (which corresponds to Aristotle's discussion on future contingency in [3, *PeriH* 9]), the purpose of Aristotle is to pass on the number of categorical propositions when

---

[8]Cf. [6, p. 154], who has shown that the algorithm and the whole calculation of this number would be one of the differences between Ammonius's and Boethius's commentary, since the result of the respective calculations differs one from another. He is clearly skeptical about the importance and value of this calculus because he has presented it as an absurd count. (Ibid., p. 154). Similarly, [18, p. lxxxix], is rather inclined to consider this arithmetic idle (pp. liv-lv); and "bizarre topic (...) valueless as a contribution to logical theory". Cf. also [18, p. lxxxvii].

[9]In [9], Stephanus's commentary, there are no explanations of why this calculus is pertinent. The character of his exposition, based on brief explanations (*práxis*), can certainly explain in part this absence, but lacunae in relevant parts must be considered. At [9, p. 24, 37-p. 25, 39], Stephanus presents the calculus of the number of the two elements propositions (e.g. 'Socrates walks'), and at p. 39, 28-32, the calculus of the number of the three-term ones (e.g. 'Socrates is just') following the pattern already established by Ammonius of counting separately the three different syntactical groups of categorical propositions distinguished by him in the introduction of the commentary: assertoric with two and three terms (e.g. 'Socrates walks'; 'Socrates is just'), and modals ('Socrates speaks clearly'; 'Socrates is possibly just'). [5, p. 7, 15, p. 8, 24].

these contain definite and indefinite names.[10]

> ... now his intention is to pass on the number of categorical propositions, whatever they are compounded, in an absolute sense, by definite or indefinite name.

The point is also made by Ammonius's commentary [5, p. 8, 20-23] though by means of a stronger formulation involving not only what is our *PeriH* 10, but also the entire treatise.

> Thus, then, after having gone through each of all the species of propositions and having faced firmly the task to show that the oppositions are not more than these, Aristotle puts end to his enterprise.

According to Ammonius, after presenting the species of categorical proposition and devoting to each of them a section of *PeriH*, Aristotle finishes the treatise. He assumes that the First Section is devoted to the elements of the categorical proposition (i.e., name, verb, indefinite name, etc.); the Second Section deals with the two-term propositions (e.g., 'Socrates walks'); the Third Section covers the three-term propositions (e.g., 'Socrates is just'); and the Fourth Section includes the modal ones (e.g. 'Socrates is possibly just'). Finally, the Fifth and last section (what is our chapter 14) relate to matters introduced once the complete treatise was finished, which is the reason why Ammonius, as Porphyry before, were skeptical of its authorship [5 p. 251, 25-p. 252, 10]. Now, Aristotle's text does not say expressly that he is aiming at counting propositions, but Boethius's commentary supports this exegesis by referring to the first words of Chapter 10, namely:

> Now an affirmation signifies something about something, this last being either a name or a 'non-name'; and what is affirmed must be one thing about one thing. (...) So every affirmation will contain either a name and a verb or an indefinite name and a verb. Without a verb there will be no affirmation or negation.

---

[10]Cf. [12, 2, 20-23, p. 251]. (...) nunc haec eius intentio est, ut categoricarum propositionum numerum tradat, quaecumque cum finito vel infinito nomine simpliciter fiunt.

According to Boethius, Aristotle's starting point would be the categorical proposition that contains a subject and a verbal predicate. Boethius's sample is 'a man walks' (*homo ambulat*).[11] Aristotle would advance to a classification of this basic proposition when containing definite or indefinite names. The starting point is the subject, that is, the name; and given that every name is a name or an indefinite name, the subject will be definite or indefinite.[12] Thus, the classification that Aristotle would bear in mind at the beginning of Chapter 10 would be this: every two-term proposition holds a subject that is either a definite or an indefinite term. Boethius' scheme is as follows:

    Unquantified with simple name as subject:
    A man walks   A man does not walk

    Unquantified with indefinite name as subject:
    A not-man walks   A not-man does not walk

    Universal with simple name as subject:
    Every man walks   No man walks

    Universal with indefinite name as subject:
    Every not-man walks No not-man walks

    Particular with simple name as subject:

---

[11] Cf. [12, 2, 18-21, p. 255]. The sample of the proposition that the commentary uses to refer to its observations is 'homo ambulat', ('a man walks'), which is an example of a categorical proposition with two terms. Later, (at 19b19 f.), Aristotle will introduce another group of categorical propositions: that with three elements, or that in which 'is' is predicated additionally as a third thing, e.g., 'a man is just'. As such, these two groups are called simple propositions, because they predicate one thing from another [cf. 12, 8-9, p. 127], so that the force lies on the verb, [12, 2, 10-12, p. 105], or on the predication [12, 2, 15-18, p. 105]. For this reason, the simple proposition is also called predicative, or categorical (if one follows the Greek language: [12, 2, 13-24, p. 186], since its unity depends on the predication and not on the conjunction [12, 2, 5-6, p. 105]. When it depends on the conjunction is called hypothetical, or conditional. It must be recalled that Ammonius emphasizes too the number of words compounding these categorical or simple propositions to make a difference between these two groups above mentioned [5, p. 78, 26 - p. 79, 9]).

[12] [12, 2, 1-4, p. 256] necesse est semper categoricam propositionem aut nomen habere subiectum aut illud quod dicitur infinitum. vero nomen est quod ipse nunc innomine vocat. The obvious consequence is that the number sought out is simply that of all categorical propositions, without qualification.

> Some man walks   Some man does not walk
> Particular with indefinite name as subject:
> Some not-man walks   Some not-man does not walk

The diagram that Boethius's commentary presents [12, 2, 19-2, pp. 257-8] is a reduced one, since it leaves aside the singular propositions, but it is more complete than that introduced by Aristotle, who only lists, says the commentary, the unquantified propositions and their contradictions.[13] Boethius completes Aristotle's list by adding the particular propositions (with definite and indefinite subjects), which are called the contrary and subcontrary propositions, in a clear allusion to the Square of Opposition of Chapter 7,[14] and suggesting that the classification that his commentary gives and that offered by Aristotle are one and the same.

> The number of propositions, which we also described above, he himself [i.e., Aristotle] proposed: first, the indefinite propositions, then the opposite propositions, because of which if any comes back to that or applies his mind to this, he accurately will know how the Aristotelian disposition differs from our disposition. Indeed, we have proposed the contrary and the subcontrary propositions, but Aristotle posed only those oppositions that are contradictory between them.[15]

According to the commentary, the fact that Aristotle proposes only some of the categorical propositions in the first paragraph of Chapter 10 does

---

[13]Indeed, Boethius argues that Aristotle first listed the unquantified: 'a man is' — 'a man is not'; then: 'a not-man is' — 'a not-man is not'. After these, the quantified (of which Aristotle gives only the universal affirmative and negative, i.e., A and E, with a definite and indefinite subject), thus: 'every man is' — 'every man is not'; then: 'every not-man is' — 'every not-man is not'. [13, *PeriH* 19b15-19].

[14]i.e., 'some man is' — 'some man is not', (viz. I, O); and then 'some not-man is' — 'some not-man is not'.

[15]Cf. [12, 2, 7-14, p. 263] numerum vero propositionum, quarum nos supra quoque descripsimus, ipse [i.e., Aristotle] subiecit: indefinitas quidem prius, post vero contra iacentes. quod si quis vel ad illa revertitur vel hic intendit animum, in quo vel nostra vel Aristotelica dispositio discrepet diligenter agnoscit. nos enim et contrarias proposuimus et subcontrarias, Aristoteles vero solum contradictorie sibimet contra iacentes oppositasque proposuit.

95

not prevent that his general purpose is to embark on a general classification of these propositions. One proof that Aristotle proposes this classification would be that he lists them in a certain order: first the unquantified affirmation (with and without indefinite subject) with their contradictory negation, and then their extrinsic tenses, that is, past and future tenses.[16]

3 The three-term propositions

Along with this exegesis, Boethius wonders why Aristotle did not divide the predicate of two-term propositions into an indefinite and definite verb, and his answer is a complete doctrine that he does not mention in his first commentary on *PeriH*,[17] and Ammonius did not know in detail.[18] This doctrine states that Aristotle did not make any difference regarding the verb, because the indefinite verb negates the affirmation, which implies that in two-term propositions there is no affirmation with

---

[16] Aristotle at [13, 19b18] says that "For times other than the present the same account holds". Boethius comments on this statement [12, 2, 14-19, p. 263]: Sed Aristoteles non solum in praesenti tempore easdem propositionum dicit esse differentias quas proposuit, sed etiam in aliis quoque temporibus quae sunt extrinsecus. extrinsecus autem tempora vocat quae praeter praesens sunt praeteritum scilicet et futurum. Ammonius, however, [5, p. 90, 21- 30], takes this point as the proper division of the predicate in the two-term propositions, since here the predicate is a verb, and the verb "additionally signifies time", and time is understood in three ways: present, past, and future.

[17] As we know, Boethius devoted a double commentary to *PeriH*. At the beginning, of the Book Four of his second edition, [12, 2, 2-4, p. 251], Boethius says that the subtleties and more difficult points are left for his second edition. The points of a classification of the simple categorical propositions and that of their number as well as this point wondered here is clearly an example of what he considered more difficult and subtle.

[18] Ammonius's commentary agrees with the following explanation given by Boethius's commentary that there are no indefinite verbs in propositions, but the indefinite verb states the negation, because both follow an explanation given by Alexander of Aphrodisias in this sense [5, p. 157, 9-24] and Boethius [12, 2, 8-16, p. 317]. However, Ammonius rather agrees with the idea of dividing the predicate, since this is always a verb in two-term propositions, and the verb holds its proper temporal three-fold division (present, past, and future).

an indefinite verb.[19] The doctrine seems to be a way to introduce a second group of categorical propositions in which the predicate can be definite or indefinite. Aristotle's own formula says [3, 19b19], those in which 'is' is predicated additionally as a third term.[20] Accordingly, Aristotle (19b27-29) will determine the general classification of these propositions through their oppositions:

(a) a man is just   (b) a man is not just
(c) a man is not-just (d) a man is not not-just

Boethius says that Aristotle first (*nunc*) makes it manifest "the order subjected to number" (*ordo subiectus numerum*), and later (*posterius*) he shows that the three-term propositions (which are unquantified) can also be considered as universal or particular.[21] Again, if the singular propositions are added, which Boethius does not consider but recognizes as the fourth species,[22] there will be produced a four-folding classification for

---

[19] Boethius [12, 2, 18-20, p. 258] idcirco de his reticuit, quod hae magis quae ex verbo infinito sunt ad unam qualitatem pertinent propositionis id est ad negativam. cf. [12, 2, 21-22, p. 261]. This doctrine can be traced back down to Alexander of Aphrodisias, because Ammonius [5, p. 157, 9-24] reports on it and mentions Alexander. As much as we can see, this doctrine and the general classification of the simple categorical propositions ground each other. The temptation of attributing to Alexander, consequently, the opinion that Aristotle in Chapter 10 would deal with the number of simple categorical propositions and even the foundation that Boethius's commentary makes of this, is great, but we do not know whether Alexander maintained this statement or not.

[20] Aristotle [3, 19b19] says: "But when 'is' is predicated additionally as a third thing, there are two ways of expressing opposition. (...) Because of this there will be four cases (...). Ackrill transl. [3].

[21] Boethius [13, 2, 13-17, p. 267] atque hoc [i.e., the order of 19b26-29] quidem in indefinitis. posterius autem monstrabitur hoc etiam in his esse quae determinationem habent universalitatis vel particularitatis.nunc autem horum ordo subiectus numerum oppositionemque declaret. 'Posterius' refers here to the following diagrams introduced by Aristotle where the remaining species of propositions are presented.

[22] Zimmermann [18, p. lxxxix, n. 2], has noted that Syrianus's result of 144 species of categorical propositions "is an accurate summary of propositional patterns discussed by Aristotle, except that there is no textual foundation for patterns with "indefinite" proper names like 'not-Socrates'". This is textually correct, but it is also true that Boethius's commentary offers this textual foundation at the beginning of Chapter 10, where Aristotle would propose a general classification of the propositions. Thus, the fact that Aristotle does not mention literally these patterns does not imply

each of these propositions, namely, universal, particular, unquantified, and singular propositions. Now, since they can also hold an indefinite subject as well as an indefinite predicate,[23] the final classification will contain four more branches for each of these species already established. Finally, as they can be affirmative or negative, the general classification will incorporate a two-folding division for each above-mentioned proposition and the number will be 32 species, the double of two-term propositions.

According to Boethius, this classification and calculus is implicit in the diagram developed by Aristotle at [13, *PeriH* 19b27-29] and explains what Aristotle himself has expressed later at [13, 20a1], precisely after he has distinguished the last opposition of this group, namely:

There will not be any more oppositions than these

Boethius affirms that Aristotle has counted the totality of the species of two- and three-term propositions, and he together with the ancient commentaries emphasized this phrase as a sign of Aristotle's intention of completeness in classifying categorical propositions.[24] Once this has been stated, Boethius's second commentary transfers Syrianus's arithmetical calculus of the total number of categorical propositions. Boethius relies on Syrianus for disposing of "these propositions liter-

---

that they are outside of his intention. In other words, the singular propositions are left aside by the classification of the first group of simple categorical propositions (singulare habentibus subiectum simplicibus propositionibus reiectis. Cf. [12, 2, 28-6, pp. 256-7]). However, this does not mean disregarding them as a valid species of simple categorical propositions, as is clear from the fact that the commentary includes here subjects which are universal things or singular things (huius autem quae ex finito et simplici est species sunt duae: quae aut universale nomen subicit, ut homo ambulat, aut singulare, ut Socrates ambulat. [12, 2, 8-11, p. 256]).

[23] Because of their peculiar syntactic structure, the commentary pertinently remarks that here what is principally predicated is a name or an indefinite name. ([12, 2, 2-5, p. 267]. quando est tertium adiacens praedicatur, quod principaliter praedicatur aut nomen erit aut infinitum nomen).

[24] So Ammonius [5, p. 175, 25-31]; but see his use of *téleios* at [5, p. 176, 1]. Also Boethius [12, 2, 14-16], ut autem quinta oppositio repereri possit, nulla rerum ratione possibile est ("so it is not possible to find a fifth opposition, because it is not in the nature of things".) And even Syrianus, in Boethius's report at [12, 2, p. 322, 20-21], who uses in his calculus Aristotle's expression of 20a1 (magis plures autem his, ut ipse ait, propositiones inveniri non possunt)

ally",[25] as revealing that all the species distinguished up to here have been well collected from Aristotle's *PeriH*.

Aristotle would call 'a man is', 'a man is not'[26] first affirmation and negation because they are first in an absolute sense (*simpliciter*), since they are the propositional elements from which propositions of three elements (or where 'is' is predicated additionally as a third element) are formed. Therefore, as the commentary points out, these four propositions whose diagram Aristotle builds up at 19b27-29,[27] are the result of the combinations of the 6 *simpliciter* ones, namely: (1.) a man is, (2.) a man is not, (3.) it is just, (4.) it is not just, (5.) it is not-just, (6.) it is not not-just.[28]

Ammonius also accepts this order between propositions with two and three terms, but he does not introduce the analytical detail present in Boethius.[29] The procedure used by Boethius's commentary in reducing these 6 *simpliciter* propositions to the 4 of three terms (but simple ones) of 19b27-29 evokes Aristotle in [3, *De Gen* 2, 330a30], where the combination of primitive physical elements is referred to. Here Boethius

---

[25]cf. [12, 2, 24, p. 324]. "We have disposed these propositions literally (*nominatim*), by collecting them from Syrianus's calculus, since it will be easier to have confidence in the number (*facilior fides habebitur numero*), if they are displayed by means of examples, and equally too because he who learns wrongly about these propositions, will teach them very incorrectly, and by putting affirmations instead of negations, and negations instead of affirmations, will confuse the complete order; and thus, since his proposal (*oratio illius*) would distort this order from the truth of a correct reasoning, we made this disposition as a help for a stronger memory (*ad tenacioris memoriae subsidium*)."

[26]In English, expressions such as 'a man is' are awkward. It must be taken as in 'a man walks', i.e., those in which another verb than the verb to be is predicated.

[27]Namely, 'a man is just' – 'a man is not just' (in front horizontally). Then, 'a man is not not-just' – 'a man is not-just' (below and in correspondence with the row above).

[28]Cf. [12, 2, 7-12, p. 268] hae quattuor propositiones ex senario propositionum numero ad pauciora reductae sunt. si enim simplices et ex duobus terminis fuissent, hoc modo essent: homo est, homo non est, iustus est, iustus non est, non iustus est, non iustus non est, et essent hae sex propositiones.

[29]Cf. Ammonius [5, p. 78, 26-p. 79, 9] The two-term proposition is called by Ammonius the 'simplest', and 'the most primitive' proposition, because it consists of the minimal terms forming a proposition, i.e., name and verb. Ammonius also explains that the affirmation of two terms is prior to the corresponding negation. Equally, the proposition is one, if it signifies one thing.

adds the following two statements:[30] (i) the number of propositions in which 'is' is predicated additionally as a third thing cannot produce a number of propositions greater than that produced by the propositions *simpliciter*;[31] and (ii) if 'man' is taken as subject and 'just' and 'not-just' as predicates, then only four propositions and two oppositions can be generated, as Aristotle showed in the diagram at 19b27-29 [12, 2, 5-10, p. 269].

The idea of composing new forms of categorical propositions from six primitive forms of two terms is a far-reaching idea. Not all consequences are drawn by the commentary, and so it is worth noting that: (i) every simple proposition is two- or three-term, if name and verb, indefinite name and indefinite verbs are accepted as terms, namely, indivisible logical atoms.[32] Then, (ii) there are 8 logical properties allowing the logical analysis of simple propositions: quality, quantity, modality, matter, indefinite subject, indefinite predicate, singular terms, and tense. Accordingly, (iii) the methodology presented might inchoate the idea that it is possible to find out all possible propositions by means of a rational procedure and extend our logical limits. Boethius's expression "[given certain propositions] *rationabiliter oriuntur (aliae propositiones)*" [12, 2, 2-3, p. 269] is a remarkable ancient testimony of this important idea in Renaissance and Modern Logic.

4 The arithmetical calculus

Regarding these earlier points, the presence of an arithmetical calculus in the ancient commentaries of [13, *PeriH*] cannot seem so odd now.

---

[30]The principle stated by the commentary has two formulae: (a) res simplices iunctae naturaliter redeunt pauciores, [12, 2, 18, p. 268]; i.e., "things which are simple become, by nature, fewer in number when they are put together"; and (which is equivalent) (b) coniunctio enim ipsa numerum minuit, "the conjunction itself reduces the number of things".

[31][12, 2, 11-16, p. 269] qui vero dixerunt numerosiores fieri propositiones ex his, in quibus est adiacens praedicatur, quam ex his, quae duobus terminis constarent, illos non intellexisse rerum naturam manifestum est, quae ita fert, ut semper ex pluribus simplicibus rariores redeant res paucioresque coniunctae.

[32]Namely, (i) with two elements, –name and verb– or secundum adiacere, and (ii) with three elements or tertium adiacere –or those in which 'is' is predicated additionally as a third thing.

The calculus makes sense because without identifying every type of categorical proposition and explain its origin, there will no certainty that we have all the contradictory opposites. If every possible species can be identified and classified, then the arithmetical calculus simply confirms the correct determination of the number of these species.

Now, are these calculi reliable? The algorithm and the calculation presented by Boethius and Ammonius, though different from one another, are both entirely correct. Both calculi are completely congruent with the properties of the categorical propositions that they previously admitted. The difference is because Ammonius admits a greater number of variables in his calculus. In contrast to Boethius, the Alexandrian commentator adds the modal ones and divides between the quality of the proposition and the quality of the *dictum*. According to Boethius, Syrianus starts from the unquantified proposition as a proposition type. He divides this proposition into those with two terms and with three terms. Then, he obtains the result of 6 types: in the branch of the two-term propositions, he correctly sees:

(1) the one with definite subject, and
(2) the one with indefinite subject.

Among the three-term propositions, he correctly distinguishes:

(3) the one with definite subject and definite predicate,
(4) the definite subject and indefinite predicate; then,
(5) indefinite subject and definite predicate; and
(6) indefinite subject and indefinite predicate.

Now, since there are 4 quantified propositions (Syrianus incorporates the singular proposition) these 6 types become 24, and since every proposition is affirmative or negative, the result will be 48.[33]

Ammonius makes an arithmetical count once he finishes the three

---

[33] Syrianus increases this result to 144, as Boethius's commentary remarks, if the 48 simple propositions are multiplied by the three ways in which the predicate can be related to the subject, that is, the three qualities of predication that Boethius acknowledges: qua in re quoniam tres sunt aliae qualitates propositionum, quae sunt necessariae, contingentes et inesse tantum significantes, secundum quas qualitates istae omnes propositiones proferuntur, has quadraginta octo propositiones si in ternarium numerum duxerimus, scilicet propositionum qualitates, centum quadraginta quattuor omnis propositionum praedicativarum, de quibus hoc libro tractat, numerositas

consecutive parts of his commentary: the first and second parts are devoted to the two syntactic groups of categorical propositions distinguished by Aristotle, namely, the two- and three-term propositions; the third part focuses on the modal propositions. In the first and second parts, Ammonius adopts Syrianus's algorithm, but he also adds a threefold division of tense and the threefold division of the matters of proposition (necessary, impossible, and contingent matters). For two-term propositions the number increases to 144 and the number of three-term propositions will be double: 288. But the most remarkable advance that Ammonius makes is the way in which he incorporates modality. His distinction between negating the modality and negating the *dictum* allows him to add the three modalities and their qualities (affirmative and negative).[34] Thus, 144 two-term propositions are now multiplied by 3 modalities and 2 qualities, which is 864. And the three-term non-modal propositions which are 288 become 1728, if they are multiplied by 3 modalities and by 2 qualities. If we summarize 144+288+864+1728= 3024, which is the number of all the categorical propositions that Aristotle distinguished and can be counted.

5 The source of the arithmetical calculus

F. Zimmermann [18, p. lxxxvii, and pp. liv-lv], has argued that the proposition count can only be understood as a by-product of the Alexandrian division of *PeriH* into 5 sections, for in this division the two-elements, three-elements and modal propositions would occupy the respective three main sections of the text. Every section is crowned by an arithmetical count of the corresponding proposition type. He thinks that here "The arithmetics are idle; but the proper purpose undoubtedly is to explain the rationale of the division of the text" (pp. liv-lv). But this interpretation seems to be rather incomplete. In fact, not only

---

crescet. Cf. [12, 2, 28-8, pp. 322-3]. For a description in full of these qualities see [12, 1, 20-8, pp. 105-6]. Boethius here in his effort to follow Syrianus confuses the matter of proposition and modality. Ammonius will distinguish these properties at [5, p. 88, 12-23].

[34] A modal proposition, namely, M(P), can make negative either its modality or its dictum. Thus, 'Necessary that P'/'Not necessary that P' are contradictory, but 'Necessary that P'/'Necessary that not-P' are not.

Boethius's commentary does not hold the division of *PeriH* adopted by Ammonius and referred to by Zimmermann, but also the so-called proposition count can be explained by an intrinsic reason, namely, the intention of Aristotle at Chapter 10 of proving that every affirmation has a single negation. Therefore, the analysis of Boethius's commentary shows that the proposition count is a theoretical piece solidly adjusted to the course of Chapter 10, (and for Ammonius to the rationale of the complete book of Aristotle), so that the calculus of a number appears rather be a proof of Aristotle's intention of assuring completeness to his theory of negation.

Now, Boethius follows literally Syrianus to make the numeracy, but the entire exegesis developed by Boethius in [12, Book 4] is not Syrianus's.[35] This exegetical line should come, then, from his main and current source, which is a Porphyrian copy of the commentary on [13, *PeriH*], as Shiel [14, pp. 349-372], has convincingly argued. We do not know with certainty, however, how much of this Porphyrian copy is due to Alexander of Aphrodisias, and it would be not surprising that Boethius's exegetic idea comes from him. It is not unlikely though, for Alexander was not unaware of Aristotle's interest in counting the elements of a set as said before (cf.above n.5), and Boethius also says that Alexander advanced "together with many others (Peripatetics)"[36] a substantial part of the exegesis, namely, that there is no indefinite verb in the two-term affirmation, which distinguishes between the primitive two-term propositions and the canonical "S is P" of three terms.

---

[35] [12, 2, 20-24, pp. 321-324]. "Consequently, after these things that have been presented by Aristotle, let us transfer, from Syrianus, whose nickname is 'Philoxenus', as we referred to above, the number of all the propositions that are examined in this discussion of the book [i.e., PeriH.], a subject-matter entirely appropriate and also useful".

[36] [12, 2, 8-16, p. 317]. et hoc quidem Alexander Aphrodisius arbitratur ceterique complures. idcirco enim aiunt non posse fieri ex infinito verbo adfirmationem, quoniam sicut verbum est infinitum verbum mox totam perficiet negationem, sic etiam verba quae in sese conplectuntur verbum est non facient infinitam adfirmationem, sed potius negationem. The expression "and this is considered by Alexander of Aphrodisias and many other [authors]" could refer perfectly well to commentators before Alexander. They could have developed some or most of the reasons leading to prove that there are no indefinite verbs in affirmations.

6 Conclusive Remarks

According to the most important ancient commentators of [13] Aristotle's *PeriH*, the philosopher distinguished, classified, and counted all the different species of simple or categorical propositions. He exhausted the set by a constructive method that relies on atomic propositional items of the form 'subject plus verb'. Here, as in other passages of his extensive work, he seems to seek the elements of a genus, which confirms his intellectual disposition. According to the commentators, the aim of his effort is to prove that every species of a categorical proposition has a single negation, which is central to his logic. Aristotle's *PeriH* would inspire the program when stating that "there are no more oppositions than these"[37] and "every affirmation has a single negation".[38] Accordingly, the propositional count is a logical requirement for his theory and confirms the importance of the contradictory negation for Aristotle's investigation.

We know that identifying the elements of a theory is essential to logic. Accordingly, it would be not surprising that propositional completeness plays a crucial role to prove Aristotle's general principle that every affirmation has a single negation. The count of the categorical propositions developed by Boethius's commentary implies that the three-term propositions are constructed from the two-term ones, which are non-analyzable propositional items. As Boethius, Ammonius remarks that Aristotle takes the name and the verb as the principles of the categorical proposition [5, p. 7, 16-22], but he goes even further by paralleling the elements of the categorical proposition and the axiomatic elements of Geometry [5, p. 7, 15-25].

The propositional count not only defines every species and its negation (or contradictory opposite),[39] but it also helps us to distinguish

---

[37] [13, *PeriH* 20a1].

[38] Aristotle mentions this principle four times in his *PeriH* and one time in his [3 *An Pr*] cf. [13, *PeriH* 17a33-34; 17b32; 18a8; 20b34] and [3, *An Pr* 51b33-36].

[39] Perhaps the characteristics of being 'countable' and 'ordered' can also be defended in this attempt. The terms 'complete' (and 'completeness'), 'countable' and 'ordered' are used here in a non-technical arithmetical sense. In particular, 'a complete set' means here 'a set containing all the possible species'; 'an ordered set' stands for 'a set with species distinguishable one from another', and 'a countable set' means 'a set with a finite number of members able to be put into one-to-one correspondence

eight logical properties of the simple proposition and their function for propositional analysis. Knowing the parts helps to understand the whole and vice versa. And thus, the eight logical properties make an obvious benefit not only to teaching logic but also to the theory, since it opens important titles to discuss in logic and philosophy of logic, along with helping us to understand our declarative language much better. It is something to discuss further, but this forgotten chapter of ancient logic suggests clearly that the history of Western logic has been a commentary on these eight logical properties.

References

[1] Barnes, J. et al., *Alexander of Aphrodisias on Aristotle's Prior Analytics 1. 1-7.* Cornell University Press, Ithaca/New York, 1991.

[2] Barnes, J. "Syllogistic in the Anon Heiberg", in *Byzantine Philosophy and its Ancient Sources.* Ierodiakonou, K. (ed.), pp. 97-137. Oxford: Oxford Univ. Press, 2002.

[3] Barnes, J. *The Complete Works of Aristotle.* The Revised Oxford Translation. J. Barnes (ed.). Vols. 1 and 2. Princeton, 1991.

[4] Blank, D., *Ammonius on Aristotle On Interpretation 1-8.* Translation with notes, in *Ancient Commentators of Aristotle*, R. Sorabji (Ed.), London, 1996.

[5] Busse, A. *Ammonii In Aristotelis De Interpretatione Commentarius,* A. Busse (Ed.), in *Commentaria in Aristotelem Graeca,* vol. iv, 4.6, Berlin, 1895.

[6] Chadwick, H. *Boethius. The Consolation of Music, Logic, Theology and Philosophy,* Oxford, 1981.

[7] Correia, M. "Categorical Propositions and logica inventiva in Leibniz's *Dissertatio de arte combinatoria (1666)*", *Studia Leibnitiana*, 2002, 34, 2, pp. 232-240.

[8] Gerhardt, C.J. 'Leibniz, G.W., Dissertatio de arte combinatoria (1666)', in *Die philosophischen Schriften von Gottfried Wilhelm Leib-*

---

with the set of positive integers.

niz, C.J. Gerhardt (Ed.), vol. IV, pp. 27-102, Georg Olms, Hildesheim, 1960.

[9] Hayduck, M. *Stephani in Librum Aristotelis De Interpretatione Commentarium*, M. Hayduck (Ed.), in *Commentaria in Aristotelem Graeca*, vol. v. 18, 1.3, Berlin, 1885.

[10] Kalbfleisch, C. *Galeni Institutio Logica*, Leipzig, 1896.

[11] Kieffer, J.S., *Galen's Institutio Logica*, J.S. Kieffer (transl. and notes), Baltimore, 1964.

[12] Meiser, C. *Anicii Manlii Severini Boetii Commentarii in Librum Aristotelis PERI ERMHNEIAS. Prima et secunda editio.* C. Meiser (Ed.), Leipzig 1877-1880. [= *PeriH* 2,]

[13] Minio-Palluelo, L. *Aristotelis Categoriae et Liber de Interpretatione*, L. Minio-Paluello (Ed.), Oxford, Oxford University Press, 1949.

[14] Shiel, J., 'Boethius's Commentaries on Aristotle', in *Aristotle Transformed*, R. Sorabji (Ed.), London, 1990, pp. 349-372; (Text originally in *Medieval and Renaissance Studies* 4, 1958, pp. 217-44; and in *Boethius*, M. Fuhrmann und J. Gruber (Hrsg.). Wege der Forschung, Band 483, Darmstadt, 1984).

[15] Wallies, M. *Alexandri in Aristotelis Analyticorum Priorum Librum I Commentarium*, M. Wallies (Ed.), in *Commentaria in Aristotelem Graeca*, vol. 2.1, Berlin, 1883.

[16] Wallies, M. *Ammonius in Aristotelis Analyticorum Priorum Librum I Commentarium*, M. Wallies (ed.), in *Commentaria in Aristotelem Graeca*, vol. 4, 6, Berlin 1890.

[17] Wallies, M. *Ioannis Philoponi in Aristotelis Analytica Priora*, M. Wallies (Ed.) in *Commentaria in Aristotelem Graeca*, vol. 13. 1-2, Berlin, 1905.

[18] Zimmermann, F.W., *Al-Farabi's Commentary and Short Treatise on Aristotle's de Interpretatione.* Translation, introduction and notes, Oxford, 1991.

Inferring Without Truth: Revisiting some Buridan's Puzzling Inferences

Manuel Dahlquist* and Luis A. Urtubey**
*Universidad Nacional del Litoral - IHUCSO - Universidad Autónoma de Entre Ríos, Argentina
**Universidad Nacional de Córdoba, Argentina
manuel.dahlquist@gmail.com - luis.urtubey@unc.edu.ar

1 Introduction

In the first insoluble of chapter 8 of *Sophismata*, Buridan proposes that the following quotation contains a valid inference:

> (PS) First sophism: Every proposition is affirmative, therefore no proposition is negative *Primum sophisma: Omnis propositio est affirmativa; ergo, nulla propositio est negativa.* ([5]: 952)

This inference appeals to the self-reference of the conclusion to show that what we call (following [28]) the Classical Account of Validity (CCV), has a counter example, for which we should accept that there are good inferences in which the premises can be true and the conclusion false; so, the preservation of truth understood as a criterion of validity, is not reliable. How then should we justify the intuitive goodness of the inference? In order to solve this problem, Buridan proposed (both in *Sophismata* and in the *Tractatus de consequentiis*) a new criterion, alternative to truth-preservation as the account of validity:

Authors want to thank Jens Lemanski for his generosity and people skills.

(M. D.): I thank Calving Normore for believing that some of the ideas I began to work in this paper were interesting enough. To Stephen Read for his continued generosity and wisdom. To my colleagues at UNL. To my wife Pamela for her love and to my children Felipe and Olivia, because they make me think differently.

> The fifth conclusion is that for the validity of a consequence it does not suffice for it to be impossible for the antecedent to be true without the consequent, if they are formed together ... Therefore, something more is required, namely, that things cannot be as the [premise] signifies without being as the [conclusion] signifies. ([28]: 955)

Thus, for the last proposal of Buridan the validity criterion is preservation of *signifying as is the case* or signifying as things are (*significans sicut est*); Spade invents the neologism 'firm' for the property of signifying as things are, and Klima calls this property, 'correspondence'. We know that Buridan's theory features an intriguing combination of semantics and pragmatics [7] and we will follow this line, from a new and original perspective. Our approach to PS is intended as an application of one of the logics based on the notion of information, more precisely on the logic of *becoming informed*, which confronts with the logic of *being informed*, specifically with respect to the so-called veridicality thesis, i.e., *information implies truth* [25] [2]. This perspective is closely related to *Logica Utens* that Buridan cultivated [42] and in perfect harmony with other important parts of logic from the 14th century, like *Obligationibus* or *Disputatio*. The proposed model allows leaving aside the paradoxical character of PS, incorporating PS as a valid inference, but without abandoning the classical concept of validity, so it remains faithful to Buridan's ideas at all times. For these reasons, it can be read as an alternative to the proposals of [11] and [7].

To begin with we shall say a few words regarding the framework in which this work should be read. The slogan that brings us together is that there are in old logics an enormous philosophical potential for new logic theories; but there are several alternatives to choose from when applying new formalism to ancient theories. Dov Gabbay ([8]: 426) perspicaciously distinguished among these alternatives one that they called 'the first level overarching author's project of using (importing) logical theories and formulations arising in fields of formal logic and applying them to modelling ancient concepts and argumentation'. We think that this also describes the perspective we have adopted in this paper. The approach based on justification logics will lend us a base to support the type of reasoning and argumentation stemming from the analysis of the first insolubilium or sophism in Chapter 8 of Buridan's *Sophismata*.

In order to understand it properly, we will first mention some questions about the context in which PS appears; then what is the characteristic common to all medieval sophism and finally the informational (and dynamic) characteristics of the solutions that were presented to the sophism; all of them based on Buridan's own logical account.

i) The treatment of paradoxes in the Middle Ages derives from treatises that began to be produced in the thirteen century: the *Sophismata*, *Impossibilia*, and *Insolubilia*, all of which were related to the topics in *De fallaciis*, the seventh treatise in Peter of Spain's *Summulae*, which in turn derives from Aristotle's *De Sophisticis Elenchis*, but includes some original features (see [18]). More specific works on insolubles began to appear by the early thirteenth century at the latest and continued to be produced until the end of the Middle Ages and by the fourteenth century treatises on the specific topic became a flourishing genre of logic. (see [33]: 246; [34]; [40]: 580). Sophisms, a topic that he approached in different ways throughout his intellectual life, occupy an important place in Buridan's work. However, '[h]is final view is described in the ninth and last treatise of the *Summulae*, with the independent title *Sophismata*, in a version from the mid 1350's ([34], sec.3.5). Buridan's main concern in writing the treatises that comprise the *Sophismata* was providing university students with exercises to help them improve their argumentative skills. But why does this topic (the very nature of the notion of validity) appear in a work originally meant for the instruction of students? The answer is that this kind of meta-logical interests are in accordance with the various roles of the *Sophismata*. As Pironet and Spruyt point out, 'On a basic level the sophismata served to illustrate a theory, but they were also used to test the limits of a theory' ([23], section 3). Scott makes a similar comment: 'They were used primarily, but not exclusively, for the testing and application of logical rules' ([30]: 5). Thus, the *Sophismata* provide a fertile ground for addressing meta-theoretical questions as the ones elicited by PS. At the beginning of the *Sophismata*, Buridan explores the limits of the notion of inference and the logical rules that govern it, focusing on the problems that valid inferences give rise to when one of their elements is an indirectly self-referential proposition, as is the case with *nulla propositio est negativa*. The first six insolubles in *Sophismata* (which lists 30 in total) are devoted to examining the conditions of validity of certain problematic inferences.

ii) In the beginning to Chapter 8 of *Sophismata*, Buridan declares that he is going to deal with 'propositions that are self-referential [*de propositionibus habentibus reflexionem supra seipsas*] on account of the significance of their terms' ([5]: 952). He then adds that the chapter 'contains propositions called insolubles' (*ibid*, 952). This is very important, because what unifies the interest of medieval logicians in this kind of problematic propositions (and inferences) is not the concept of truth, since we also find other semantic concepts such as validity (the one we will examine in this work), as well as concepts of a different nature (e.g. epistemic concepts). Nonetheless, the characteristic that is clearly present – in various and interesting ways– in each and every insoluble is (some form) of self-reference.

It bear that, in the case of PS, this self-reference is of an indirect sort. As noted by Read ([28]: 193), this happens when the propositions that make up the argument contain an expression referring to a piece of discourse of which that proposition is a part. We have tried to capture this fact by introducing the notion of a (informational) scenario, where the inference occurs. A kind of stage-setting for playing the inference. In the sophism one can say that the expression 'affirmative' displays a scenario (of affirmative propositions), which involves this expression and which excludes all negative ones. By this side, the paradox also connects with the well known paradox of properties, akin to the set-theoretic paradoxes.

iii) The medieval account of logic is inextricably linked to practice; 'Buridan holds that logic is ultimately a practical rather than a speculative discipline' ([42]: *section* 4). The Logic is *Logica Utens*, and the *Logica Utens* is a theory for practical purposes: to equip disputants.The disputes were presented in the treaties called *Obli*-

> *gationes*, which provided rules that governed the disputes. This seems to fit perfectly with contemporary proposals that point out: 'in recent decades, a much broader view has emerged where logic is about a wide spectrum of common sense reasoning activities, often intertwined with natural language' ([36]: 130). This new perspective on logic is analogous to the medieval perspective.

The disputes were canonically developed as follows:

> In the most typical variations of the technique, the sequence would begin with a special proposition, called the positum. It was taken as the starting point, which the rest of the sequence would develop. The respondent had to accept the positum, if it was free from contradictions. Then he had to take into account in his later evaluations of the other propositions that he must at any time during the disputation grant the positum and anything following from it. ([41]: 3)

For logicians like Buridan, the essence of logic is given by being a tool to obtain knowledge and knowledge is linked to logic, as Ashworth explains:

> [Buridan] constantly emphasizes the role of human mental processes. In part, this is due to his recognition that, as Biard put it, the production, articulation and understanding of language requires the constant intervention of thought ([3]: 201), but the main reason is his preoccupation with epistemology. *Logic is a tool used in the search for knowledge, and the acquisition of knowledge depends on our ability to identify true propositions and to use them as premises in valid arguments.* ([3]: 259; emphasis added)

It is for this reason that medieval logicians are interested in dynamic processes; dynamic in the same sense that contemporary programs express it today.

> Logic is often taken to be about propositions, truth, and proofs: abstract objects in Heaven, and their Platonic properties. But the discipline arose in Antiquity by studying activities on Earth: dialogue and argumentation. And the

> very terminology of logic still has a double meaning. 'Statement' is both a dynamic activity and the static product of that activity, 'proof is a procedure of establishing a claim and a formal record of that procedure, etc. These activities are usually kept in the background, as mainly didactical motivation. Placing the dynamics at centre stage in logical theory is the program of 'logical dynamics'. ([37]: 183)

But also, and especially in belief–dynamic processes:

> First of all, a disputation is a dynamic process driven by 'epistemic input' in the form of incoming sentences to be incorporated into a larger body of sentences. *Secondly, that the first sentence (the positum) in a disputation should always be accepted corresponds to the so-called success postulate in belief revision theory.* Third, the central goal in both frameworks is to avoid inconsistency, and, fourth, there is also an inherent conservativity or minimal change principle at work in both cases. ([14]: 35; emphasis added)

This method over time -and this is historically relevant data- was used to deal with paradoxes: 'The link between insolubles and obligations was established especially by authors who, like William Heytesbury (before 1313-1372), solved the problem of insolubles in terms of obligations ([24]: 98). The most important thing is that, by adopting this method, the solution to the paradoxes acquired a contextual imprint, since the function that the positium fulfilled was to generate the context or -as we will say here- the 'stage-setting' where the paradox happens.

> *No doubt circularity sometimes occurs by chance but in general we need to arrange the circumstances of utterance, inscription, or thought so as to guarantee it. This stage-setting was done in the middle ages by means of a hypothesis or positio.* The theory of such hypotheses was regimented in treatises on the obligatio of the same name and it is to these, we suggest, that one should look to understand how the mediaeval concern with insolubilia is originated and the character of the earliest solution. ([15]: 63; emphasis added)

This medieval logic account (dynamic and contextual) has consequences on the issues it addresses and on the way it addresses those issues. The idea of stage-setting (embodied by the positio) that should be understood not as a state of affairs to which the propositions refer, but as propositions that an agent utters, will be central in our approach to PS.

2 Puzzling inferences, self-reference and logic

We will present now the problem Buridan encounters in assuming the validity of PS and then the framework in which we will treat this puzzling inference. To begin with, we must not forget that we are dealing with an argument, as opposed to other sophisms which consist of only a proposition , e. g., the liar. PS is made up of two categorical propositions: 'NO PROPOSITION IS NEGATIVE' -the premise- which we will call NPN, and 'EVERY PROPOSITION IS AFFIRMATIVE', the conclusion, which we will call OPA, (by his initials in latin: Omnis propositio est affirmativa). They form the following inference: Every proposition is affirmative; therefore, no proposition is negative [*Omnis propositio est affirmativa; ergo, nulla propositio est negativa.*] ([5]: 952). The premise –OPA– is possible, i. e., we can assume it to be true, since it is not impossible that all of the propositiones that exist at a given moment are affirmative. Buridan claims that 'Every proposition is affirmative' would be true if God annihilated all negatives, and then the consequent would not be true, for it would not be ([5], 0.2.1, 953). On the other hand, NPN is always false. Roughly, Buridan argues that NPN cannot be true, since in order to be true it must exist, and if it exists, then there is at least one negative proposition, namely, NPN itself.

What [28] calls Classical Conception of Validity (CCV) and Hughes calls Theory A, 'states that an argument is valid if and only if it is impossible for its premises to be true while its conclusion is false' [10]. Such is the definition of logical consequence that follows from adopting truth-preservation as the criterion of validity. In other words, a necessary and sufficient condition for validity is that the premises cannot be true while the conclusion is false. And this was indeed the dominant view in the fourteenth century. In *Tractatus de Consequentiis*, Buridan gives his definition of consequence (and then introduces two additional definitions meant to improve upon it):

> Hence, many say that of two propositions one is antecedent to the other if it is impossible for the one to be true without the other being true, and one is consequent to the other if it is impossible for the one not to be true when the other is true, so that every proposition is antecedent to every other proposition for which it is impossible for it to be true without the other being true. ([6]: 67).

But it happens that PS does not seem to behave according to the criterion proposed by CCV, since, if we accept the truth of *Omnis propositio est affirmativa*, it follows that it should be true Nulla *propositio est negativa*, but *Nulla propositio est negativa*, is always false. So we have an argument with a true premise and a false conclusion, which, however, we are willing to accept as valid. Buridan puts it clearly:

> 0.2. Again, a consequence is not valid if the antecedent can be true without the truth of the consequent. But this is the case here, for since the antecedent can be true and the consequent cannot be true, it is clear that the antecedent can be true without the truth of the consequent. ([6]: 67).

Roughly, Buridan raises this problem: if CCV provides an adequate criterion of validity, then PS is invalid; but PS is valid; therefore CCV does not provide an adequate criterion of validity.[1] According to Hughes, Buridan argues that Theory A (or CCV) cannot be correct, and then proceeds to use 'the established validity of the sophism as a reason for rejecting Theory A' ([10]: 82). Thus, it turns out that the point of PS was to test the adequacy of CCV.

Now we will turn to our proposal, although for now we will not do so in that formal terms. Later on, we will introduce more specific formal languages to expose our solution. Moreover, for now the notion of a stage-setting just mentioned will be kept informal.

---

[1] For Buridan, PS is a valid argument and the validity of PS is established by the following three arguments: (P.1) by the *locus from contraries*; (P.2) by the possibility of generating an enthymeme ; (P.3) by the rule of Contraposition. (See [5]: 952; [10]: 80-81). Precisely, we are going to base our formalization on these postulates, by accommodating the supporting arguments in the setting of justification logics.

In PS it is said initially that in the stage-setting that presents the inference (the positio) we only have affirmative propositions (*Omnis propositio est affirmativa*) and the agents becomes informed about that. To say that the relation of logical consequence does not preserve truth in this scenario, is tantamount to say that the algorithm for carrying out inferences in a formal system does not therein guarantee the preservation of this property. Having self-reference in the language, the behaviour of this algorithm becomes a little tricky. Accordingly, to derive the paradox that Buridan obtains, we can follow the strategy sketched by [16] to get the liar sentence, who in turn inspires his procedure on the way to reach the well-known Gödel's sentence. Let α be 'Every sentence is affirmative'. As in Maudlin version, the predicate P(x) stands for '(x) is recognized as certainly true by the system'. In the case that the system can be reduced to an algorithm, P(x) can specify this fact. In the same vein, but confined to the actual scenario, we have to add a predicate Neg(x), which stands for '(x) is a negative sentence'. Considering that it can be effectively determined in the formal system when a sentence is negative, it follows that it can be always recognized by the system whether it is the case that a sentence is negative. Moreover, a 'T-axiom' is admitted: If 'S' is true, then S and also a 'Trust rule': The inference from P(**n**) to **n** is valid.

Admittedly, we have now P(T) and then that any sentence is affirmative, then no sentence is negative. A direct inference gives that no sentence is negative (denote the sentence with NPN). Again, it yields P(NPN) and so NPN is recognized as true. It entails P(Neg(NPN)) and hence Neg(NPN) is recognized by the system as true as well. Consequently, we have shown in this scenario that there is an x such that x is a negative sentence, contradicting NPN. But this last fact is not validated by the relation of logical consequence. On the contrary, according to what is ruled by logical consequence as truth-preservation, it must be true the contradictory of this outcome. Therefore, there is a sentence now in the language, recognized as true by the system, which poses a conflict with the relation of logical consequence as truth preservation. Admitting the predicates P(x) and Neg(x), the axiom T, the rule **Trust** and the consistency of the system, this sentence is proved to exist. Moreover, since NPN is true, this last consequence should be false. It is worth to observe that there could be a trouble with respect to the rule **Trust**. This rule implies –as Maudlin says– 'being a cavalier about the use/mention distinction' , in spite of the fact, as he adds, that it is done 'in an obvious way'. Consequently, it seem to be harmless here. Moreover, this device allows to dispense with the use of more cumbersome artefacts, like Gödel numbers, to talk about the syntax into the same syntax.

3 Token based semantics, evidence and truth

As many authors have noted, 'propositions are, for the medievals, concrete token utterances, whether spoken, written, or mental. Different tokens of the same type can have different truth-values' ([29]: 280). On the other hand, NPN (and other propositions of the same type) describe characteristics that apply to themselves, such as 'being negative' or 'being affirmative'. That is, they can describe facts about their own syntax without confusing the object language and the meta-language (see [35]: 495). Notably, this particular feature of the semantics can be exploited in our approach, since it allows that every proposition may count as evidence for itself; the term used in the literature to describe this property is 'connote'. We can see, then, how a sentence can connote the set to which it belongs. 'Some propositions are affirmative' is evidence of the fact that there exist affirmative propositions.[2] In this way, we can allow that the token, considered as evidence, fulfils the task of informing the semantic content of the proposition. This gives us the representation that models the stage-setting where the inference takes place. Our concern is with these special inferences considered by Buridan and other medieval logicians, which involve particularly self-referential sentences,

---

[2]Considering Prior's analysis of the sophism S. Uckelman comments: 'Each term is associated with a particular group of shapes, which it may be said to connote, though this means no more than that the presence on a sheet of marks of certain shapes will determine [...] whether or not sentences containing certain terms are to be counted as 'true on their sheets'. ([26]: 483 Definition 3.5) The connotation of a term is defined as follows: The term 'propositio' connotes all L-sentences. The term 'negativa' connotes all L-sentences whose sign of quantity is 'nulla' or whose copula is 'non est'. The term 'affirmativa' connotes all L-sentences which are not connoted by the term 'negativa'. ([35]: 491). Arguably, Prior's solution also introduces a sort of contextualization for the notion of truth.

and where they have noted that it does not hold the criterion of the preservation of truth to define validity. As Klima has remarked though,

> ... since this revision of the definition of the validity of a consequence had to be introduced only because of the possibility of a proposition token quantifying over itself in a natural language, once one keeps this possibility in mind the definition of validity need not be totally overhauled, as Buridan himself recognized, and he used the definition based on the idea of necessary truth-preservation (i.e. on the idea that the truth of the antecedent is 'preserved' in the truth of the consequent) without further ado concerning consequences not involving such self-referential propositions. ([12]: 321).

Moreover, when proposition like NPN are involved, we have a problem with the truth-preservation as a criterion of validity. It is a remarkable fact that these inferences -like PS- have to do with propositions that confirm or refute themselves. What is necessary to make 'some proposition is affirmative' true is that there exist some affirmative proposition. This requirement is satisfied by the proposition itself. On the contrary, in the case of NPN, the same proposition gives the conditions to make it false. What is required to make false NPN is precisely that there is a negative proposition, namely, itself.

> If propositions are atemporal, they exist timelessly, that is, there is no time at which they do not exist. So, whenever a timeless proposition is expressed by a temporally occurring sentence-token, then the proposition expressed by that sentence-token exists. Therefore, whenever I form a token of the sentence 'No proposition is negative' the proposition that no proposition is negative expressed by this sentence-token exists. But its existence entails that some proposition is negative, so the proposition cannot be true. ([11]: 98)

Strictly speaking they are not self-referring in the current sense of the word, but for a more dubious anomaly concerning some interference that have with their own truth-value. They reject the value they should get in these circumstances, according to the values of other propositions related with them.

4 Preserving properties other than truth

In SP inference, OPA and NPN are related; not only do they make up an inference, but It is still clear enough that premise and conclusion have the same meaning. To see that, as Sara Uckelman notes, it suffices to interchange the quantifiers according to well known equivalences in first-order logic ([35]: 489). This is fine, but it should be noted that interdefinability is not enough; justification is not complete without accepting that there are contrary predicates, i. e., pairs of predicates such that if one is a predicate of one thing, the other is excluded from that possibility; only then can we accept that 'Non-Affirmative' is equivalent to 'Negative'; for this reason Buridan appeals to the *locus of contraries* to found the validity of PS: [3]

> This is proved first by the locus from contraries. For just as 'Every man is ill; therefore, no man is healthy' is valid because it is impossible for the same [person] to be both ill and healthy, so is the above, because it is impossible for the same proposition to be both affirmative and negative. ([5]: 952).

But then and this is what Buridan faced, if one keeps the truth preservation criterion as a necessary condition for validity, one can derive these three mutually inconsistent statements:

1. There is a false proposition which principally signifies as things are, ('No proposition is negative' or NPN);

---

[3]The term 'contraries' can be applied to a relation between propositions or to a relation between terms that make up the propositions. When he talks about propositions, laws such as the following are inferred for him: 'AeB excludes the truth of AaB, since AeB and AaB are contraries, i.e. they cannot both be true together' . ([22]: 139). 'Contraries', when referring to a relationship between terms (as is the case here), it refers to a predicative relationship. 'Propositions involving different kinds of predicative relation and different kinds of subject matter will necessitate different kinds of advice; the basic unit of advice is referred to as a topic, or, in the Latin, 'a locus' ([39]: 122). Thus, every *locus* is the expression of a predicative relationship between terms of two or more propositions, with respect to some subject matter. When the locus is the *locus of contraries*, these two terms cannot both be true together, as is the case of 'affirmative' and ' negative'.

2. There is a formally valid inference with true premises and false conclusion: Every proposition is affirmative; therefore, no proposition is negative (OPA; therefore NPN);

3. There is a pair of equivalent propositions (whose *significata* are equivalent) that have different truth-values.

Buridan -like other logicians of the XIVth century- reformulates the requirement for the validity of a consequence for not agreeing with truth preservation as a criterion of validity.

> Therefore, some give a different definition, saying that one proposition is antecedent to another, which is such that it is impossible for things to be altogether as it signifies unless they are altogether as the other signifies, when they are proposed together. ([6]: 67)

Buridan introduced a new criterion for validity, one that is not based on truth-preservation but on what [32] calls *firmness*, and Klima *correspondence*: '*Buridan reformulates the requirement for the validity of a consequence in terms of the correspondence-conditions* of the propositions it involves'.[4] Preservation of firmness as the generic criterion for validity is also found in some authors in the English tradition, e.g., in Strode:

> A consequence is said to be valid (bona) when things cannot be as is exactly signified by the premises unless they are as is exactly signified by the conclusion ...... But a consequence is said to be valid in two ways: for some consequences are said to be valid in form (*bona de forma*) and some valid only in matter (*bona de materia*). A consequence is said to be valid in form when, if the way things are exactly signified by the premises is understood, the way things are exactly signified by the conclusion is also understood ...... and so it is said that in such consequences the conclusion is formally understood in the premises.[5]

---

[4]([12]: 320–321; emphasis added.)
[5]Radulphus Strode, *De consequentiis*, 1.1.02–03 ([31], quoted by [29]: 286).

It is necessary then -for the sake of the argument- that we make some clarifications on truth-preservation and preservation of other properties in the relation of consequence that seems to bear on these inferences. What happens is that logical consequence guarantees the preservation of semantic information. It also happens with logical proofs, since this preservation concerns the property of rules supported by its soundness. That is, it is a matter of soundness and not of completeness, in terms of contemporary logic. What we are saying is that whenever a consequence is settled, for example through a set of inference rules, this property is preserved, but it does not mean that all inferences preserving this property are guaranteed by a set of rules.

This is one way of understanding what Buridan does when he reformulates the requirement for the validity of a consequence in terms of the correspondence-conditions of the propositions it involves. In the *Tractatus de Consequentiis*, Buridan proposes to analyse the inference: No proposition is negative; therefore, no ass is running ([6]: 67); and understand that this definition –the definition in terms of firmness or correspondence– now guarantees that even if the antecedent automatically falsifies itself whenever it is formed, its self-falsification does not automatically validate the consequence, for it still leaves open the possibility that the situation signified by the antecedent holds without that signified by the consequent.[12].

Putting all this in cognitive terms, we can say that the semantic information given by the antecedent or premise of the inference can be separated from that given by the consequent, even though the propositions are saying the same from the logical point of view. We can countenance the information given by one proposition, separating this information from the semantic content of the other proposition. It is not excluded altogether the possibility that both hold, the situation or the stage-setting informed by the premise and that informed by the conclusion. In a stage-setting like this, when one becomes informed of the first one becomes informed of the second. Undoubtedly, in PS, one becomes informed that there are no negative propositions. Thus, it is impossible that a proposition signifies in a different way than the other. Additionally, no evidence that there is a negative proposition is considered or given by the inference, since PS's evidence, let call it s, could just be afforded by NPN itself. Thus, in this stage-setting, this evidence *s*, even though implicit and self-referential, may be omitted by the agent, because it does not take part in the way that premise and conclusion signify when they are put together. In other words, they do not 'refer' primarily to this fact or do not 'inform' about that. From the perspective we have chosen it happens something similar, for the agent sets in this stage-setting can be aware of the positive information and disregards the negative one about this topic. We can talk of meaning-awareness in this case, as long as we are taking account of something that is signified or fail to be signified by the premise and the conclusion.

5   An Information-oriented approach

It should be clear at this point that there will be alternative to deal with PS (and any arguments of that same type). One of them -the most radical alternative- would be to change the logic and so the logical consequence relation. This radical position is the one that Klima attributes to Buridan:

> So Buridan' s logic, construed strictly as the theory of validity of consequences, could in principle dispense with the notion of truth altogether. ([11]: 105)

In coincidence with Jennifer Asworth and Catarina Dutilh Novaes's interpenetration of Buridan's proposal, we are going to elaborate on a less radical departure, based on the consideration of the logical inferences that an agent can perform in a stage-setting like this. That is to say, a stage-setting where an agent confronts an inference with a proposition that *interferes* with its own truth-value.

Our approach to PS is intended as an application of one of the logics based on the notion of information, more precisely on the logic of 'becoming informed' , which confronts with the logic of 'being informed' with respect to the so-called 'veridicality thesis'. This thesis endorsed by some logics based on the notion of information explicitly relates information with truth by attributing truthfulness to information data. On the contrary, the logic of being informed rejects the veridicality thesis, thus sets aside the requirement that a proposition ϕ must be true in order to admit that you are informed that ϕ. Consequently, information of ϕ just provides assertion conditions for the truth of a proposition ϕ, provided that the agent A verifies ϕ.

Effectively, it is pointed out that '[t]he informal idea behind the logic of becoming informed is that a basic distinctive property of information is that it acquires its truth value when and if it can be verified' ([25]: 444). Moreover, with the purpose of connecting this logic with the logic of justifications, it has also been remarked that a logic for information defined on the well-foundedness of epistemic states is feasible in terms of the logic of explicit justifications (JL) ([25]: 442). Definitively, it is a remarkable point, since the logic of justification developed in the last decades, provides a useful framework in which inferences can be considered in a more brother context. Admitting more elements that take part in the course of reasoning, it can be couched in a richer formalization, where more elements are related in an argument. These other elements give the chance to include also contextual dependencies among terms and relations, which are not strictly formal or even better plainly material that link premises and conclusion. The inclusion of justification terms in the logic of justification, which functions as tags attached to propositional variables, provides a device to insert different kinds of evidence, including information concerning the truth of the proposition so attached. This is the perspective from which we want to confront this baffling inference made up by Buridan. We think that it can be considered from a perspective that incorporates justifications and a concept of consequence that also assumes its inclusion. It involves a complex inference setting including many aspects that escape to a single approach, either truth-theoretic or proof-theoretic in essence. The hypothesis about the complexity of this inference is based on the fact that it includes both contextual and material elements that make fail a purely formal approach. At the end, it will turn out that tricky inferences as that devised by Buridan, makes clear whether any single concept of consequence can be defied by common reasoning and specially when material elements are brought into the argument. What is needed is a more flexible scheme in which a diversity of components could be in principle accommodated. Assuming a broader insight, such an scheme can be proportioned by the logic of justification in the present case. We are not claiming though that the inference in question deserves a special type of consequence. Though intuitively valid, the classical model-theoretic approach cannot fully account for that. This seems to be the perplexing fact. It can be admitted that the argument exceeds this framework. We are pointing out now that it is a richer argument and that is the reason why it must be considered from a perspective that makes viable to integrate more elements of such a diverse origin. While not involving knowledge but information it is still a matter of incremental inference. Controversially, knowledge rests out of this picture, since it is a matter of truth and we have an inference without truth precisely. It makes no sense to talk about a premise guaranteeing the truth of a forever false conclusion. Admittedly, in some sense the premise is still guaranteeing something about the conclusion though. Not its truth evidently. Considering this argument as an exponent of argumentation without truth is part and parcel of a broader consideration of arguments from a logical point of view, in which the truth itself can enter the scene and go away at certain points. Anyway when truth disappears arguments can still remain sustained by other resources that are not linked necessarily to the truth or falsity of the propositions. Commenting on a thesis by Klima, [3] has made the following observation:

> Klima has argued that Buridan needs only the satisfaction of correspondence conditions, and so has a logic without truth, writing 'Buridan's logic does not have and does not need a definition of truth' (...). This may be so, but Buridan was clearly quite happy to speak of truth in accordance with conventional language.

Precisely, it is a nice feature of our approach that truth does not altogether vanished from inference, but truth is fully recovered when it gets on well with the inferential process. Another observation that Ashworth has made with respect to Buridan's logic is that it could be considered as psychologistic in a moderate sense. Ashworth writes:

> One may conclude ... that, while Buridan avoids Frege's criticisms, he does seem to be an adherent of what Susan Haack (...) calls weak psychologism: he believes that logic prescribes how we should think, but certainly not that it involves only a description of how we do think.

According to [9] strong psychologism is the position that 'logic is descriptive of mental processes (it describes how we do, or perhaps, how we must, think).' She describes weak psychologism as the position that 'logic is prescriptive of mental processes (it prescribes how we should think)'.

Our reconstruction or modelization of the inference corresponding to the sophism PS aims to see logic as prescriptive of the corresponding mental process, in the preceding sense. After defining an adequate framework, our purpose is to extend the use of justifications proposed for JL, in order to include justifications that impose certain constraints on the truth of the propositional content conferred by a sentence. This approach allows a modal interpretation where the box operator translates assertions and in which certain construction **c** counts as a justification for the truth of the propositional content **C**, that is, **c:C**, analogous to the interpretation of the box-operator in JL. Conspicuously, a basic form of JL will suffice to develop this formalization.

Turning back to PS, becoming informed that every proposition is affirmative, an agent A knows that it holds that every proposition is affirmative. One could also say that agent A recognises that it holds 'in the actual context'. This translates in the simplest form the idea that there is an action of updating on judgemental contexts [25]. Somehow A gets access to the justification **c** for the truth of the propositional content **C**. It does not mean that A is in possession of this justification. Officially, information just gives to A the 'assertion conditions' for that.

We will not enter into the details about these 'assertion conditions'. We will just assume that these are given by the information furnished by the propositional content at stake. Thus it is likely to interpret the preceding expression **c:C** in the sense that the agent A becomes informed that **c** is a justification of the truth of the propositional content **C**. It means that A is in position of asserting that **c:C**, but she could not have the justification **c**. Simplifying, we will interpret the expression **c:C** as saying that one becomes informed that **c** is a justification of the truth of the propositional content **C**. Instead of a plain justification logic we are going to retreat ourselves to a justification logic of becoming informed. That is to say that we are thinking on a logic of becoming informed on the basis of some justification for being informed about something.

Turning back again to PS, let 'every proposition is affirmative' be represented by ϕ, then there is **c**, which will count as a justification for the propositional content **C** of ϕ. That is to say that one becomes informed that things are as it is (actually) signified by ϕ. This is a plausible interpretation for the fact that things are as it is signified by the proposition, since it can be said that somehow one becomes informed that there is a justification in virtue of which the propositional content **C** of ϕ is true.

One of the outcomes of our reconstruction of the inference is that the *locus from contraries* is relegated to the composition of evidence performed by operations on terms assigned to the premises. It is not a rule that operates on the propositions themselves. It looks right since we manipulate the rule to perform a task on the evidence that supports the premises and lead them to the conclusion. Illegitimately, it might count as an extra premise of the argument, that is not part of it either explicit or implicit. At the end, is this rule that serve to manipulate evidence that will allow it to pass from the premise to the conclusion fixing the setting to validate this inference. That means that this kind of inferences we have to encode more information on the inference than that simply conveyed by the sole propositions. Among this, for example, is placed the information that the proposition has been formed! We have considered that the inference holds to textual evidence introduced by the *locus from contraries* that is part and parcel of the informational context.

6 A minimal JL of becoming informed for PS

Commenting in passing Prior's article and Buridan's notion of consequence, Dutilh Novaes observes:

> ... one of his conclusions was that it cannot be correctly formulated in terms of the notions of impossibly/possibly true, since some criterion other than their truth-values is needed to establish the (actual) modal value of propositions (with respect to each other) in order to assess the validity of the consequence relation between them. ([7]: 278)

She also adds that 'Prior's main idea was that it was necessary to differentiate the situation in which a proposition is true or false (namely, a situation in which it is actually formed) from the situation of which a proposition is true or false (namely, a situation in which it is not necessarily formed)'. This passage characterizes very well a solution to the problem stated by Buridan. In a similar vein we are going to give an alternative modal treatment of these problematic inferences appealing to a much broader inferential setting. Our coincidence with respect to the preceding quotes concerns the observation relative to the insufficiency of the criterion offered by the propositional truth-values to account for the modal status of these propositions and the logical relations between them. Moreover, from our point of view, it can be admitted for these inferences a departure from the standard view of consequence in an even more crucial way, since they are worth to be considered in a more complex setting, perhaps not strictly logical in the standard way of talking about the subject, but still inside the limits of logical modelization. In this respect, the inference in question seems to show a conflict between a theoretical notion of consequence and arguments in the real world. The real question would have to do with the application of the notion of consequence to take account of the intuition that one sentence follows from another.[6]

To formalize the type of stage-setting we want to appeal to, one needs specific domain-dependent models, with additional features ([2]. These are the kind of models introduced by S. Artemov in the setting of Justification Logic. In justification models, justifications are primary objects, and a distinction is made between accepted and other types of justifications.[7] By adopting justification logic, we are also taking advantage of the existence of 'accepted justifications' which do not produce other related effects. In the case of material inference, the status of

---

[6]It is worth noting that Buridan's solution to the insoluble should not be confused with the answer to the objections. In the first sophistry, one of the answers to the objections (the answer to objection 1) has given rise to much logical and philosophical literature. But these works (Prior's paper among them) do not deal with (nor develop) Buridan's solution to the validity of PS; they only work with the answer (very original indeed) that Buridan gives to that particular objection. We have based our treatment on Buridan's solution in this paper not on the answer to the objections that he formulates there.

[7]Specifically, in the case of Artemov's *justification awareness models* a distinction is made between accepted and knowledge-producing justifications. What is most important for us here is that in these models those concomitant effects can be controlled.

supporting justifications becomes important, since the premise which is added to perform the inference is otherwise justified. Anyone unaware of the convenience of this premise to the inference, could not introduce it into the argument. Considering the inference as a stated fact, we need to know how to represent several pieces of evidence concerning this fact, which are not explicitly included, but which somehow follow or are justified from the stated facts.

We are now ready to begin our consideration of the PS (Every proposition is affirmative; therefore, no proposition is negative). There is a peculiarity that we have to pay attention to in the formalization of these inferences. One must be aware that contrary to the usual soundness and completeness analysis of proof systems, what is needed in this case is the introduction of specific domain-dependent models, with additional features that are not necessary in the other case. Let us consider this time both inferences introduced by Buridan.

1. Any proposition is affirmative, thus no proposition is negative.

2. Any syllable has many letters, thus no syllable has only a single letter. ([5]: 952)

Contrarily, in the last case, by reflecting upon the conclusion it follows that there is a syllable that has a single letter. It is justified in virtue of the principle which can be roughly stated as follows: 'all evidence obtained from a stated fact has to be also (admitted) as a stated fact'. An analogous principle can also hold for meaning.

Following the key idea in our interpretation, we have that two models also justify these inferences. Accommodating Buridan's inference in this interpretation, it is not meant that these models correspond to something actual. On the contrary, it happens that the self-referencing propositions involved in both cases, respectively, do not exist. We just have to think that they are two models that depict the way things could happen if they were to happen at all. The inference is validated without relying on the truth and the existence, consequently, of the propositions involved. We are just considering information and not truth.

The aim of justification models is to reach more expressive power to capture differences between justifications and their consequent use by the reasoner. We have said before that some justifications can produce and other may fail to produce the desired outcome. It is the reasoner who decides which one could serve as a base for his beliefs and which justifications would be ignored. These are the kind of actions that make up our inferential scenarios. Let us focus on the central points of Buridan's example. The cases at point can be dubbed *stage-settings with self-falsifying propositions*. Concretely, this is one of the cases that Duthil Novaes [7] referred to as 'propositions interfering' with their own truth-value. The situation is the following. The agent or reasoner is placed in a stage-setting wherein one proposition (NPM) follows from another (OPA) (the positum) and it is questioned whether the inference is valid. Moreover, the more complete description of the scenario could include other features, such as:

- There are justifications s (that stems from the assumption that all propositions are affirmative) and t (the proof based on the *locus from contraries*), which are combined to justify NPN.

- Evidently, the combination of s and t does not support the truth of the conclusion, for the reason that the target proposition cannot be true along with the premise.

- Judging s and t both supply strong evidence, agents can accept NPN based on them.

- The resulting belief or becoming-informed outcome is evidence-based, without it being true.

Consequently, inference can be saved on the basis of these justifications in a scenario that validates the inference. According to our interpretation, through the evidence or justification one becomes informed about something φ. Moreover if φ implies ψ with the concurrence of another piece of evidence, then ψ also can hold supported by the combination of the evidence at stake.

7 A language to accommodate this inference

Commenting on the first Buridan's sophism, Sara Uckelman points out that it can not be treated merely as an inference in first-order quantification logic. Buridan's formulation is couched in terms of Aristotelian logic and it is well known, concerning Aristotelian and modern logic, that one of the central ideas of modern logic is to take the Aristotelian A-form sentence not as primitive, but as analysable into quite independent logical components: universal quantification and the conditional. In consequence, this is is not the more convenient approach confronting Buridan's inference for us. It seems much better to take the Aristotelian forms as primitives and to remain faithful to the subject-predicate form. [38] gives some additional clues to proceed in this line:

> It is well-known that once we have full monadic predicate logic, we have full propositional logic as well. Indeed, syllogistic can be viewed as a fragment of propositional logic. ([38]: 85-108)

This quotation fits perfectly with the ideas of medieval logicians, because categorical propositions are -by tradition- the primordial units of language since they start from the Aristotelian definition of categorical proposition.[8] A categorical proposition is one that has a subject, a copula and a predicate as its main parts, for instance, 'A man is an animal' ([21]: 124). The usual categorical ones ('Socrates is white', for example) will be semantically decoded through (semantic) categories of propositional logic, whose influence grows until founding the theory of consequence, the neural center of 14th century logic ([18]: 29-33). On the other hand, a categorical sentence is always one of the following four kinds: a-type ('All men are mortal'), i-type ('Some men are philosophers'), e-type ('No philosophers are rich'), or o-type ('Some men are not philosophers'). In a- and e-sentences the predicate is affirmed or denied of the whole of the subject, while in i and o-sentences, the predicate is affirmed or denied of only part of the subject. The syntax of categorical sentences can be formulated as follows: If S and P are primitive terms, SaP, SiP, SeP, and SoP are categorical sentences. Because a syllogism has two categorical sentences as premises and one as the conclusion, every syllogism involves only three terms, each of which appears in two of the statements.

---

[8]Aristotle defines categorical sentences as those sentences that say something about something (*De interpretatione* 5 17a). Thus, categorical propositions are -by tradition- the primordial units of language.

Considering that, in our approach, arguments in the stage-setting will be framed into the traditional syllogistic reasoning, wherein a categorical sentence is derived as conclusion from other categorical sentences, it will suffice to display the syntax of categorical sentences in the basic part of the language. Although they are controversial in some respects it is worthwhile to appeal to some propositional formulation of syllogistic language in order to take account of the argument.[9]

Customarily, a language of this type is used in presentations of axiomatic systems of syllogistic, like [13]. This language counts with name variables S, P, M, N, ..., propositional operators and the two primitive operators a and i specific for syllogistic. SaP is read as 'every S is P' and SiP as 'some Ss are Ps' or 'certain S is P'.

Formulas like SaP and SiP are called atoms. By using Backus–Naur notation, a formula of the language can be defined formally:

$$\alpha = \text{SaS} \mid \text{SiS} \mid \neg\alpha \mid \alpha \to \alpha$$

The usual negative syllogistic operators e and o can be defined as negations of the primitive operators:

- SeP $=_{Def} \neg SiP$,
- SoP $=_{Def} \neg SaP$.

Similarly, other Boolean connectives can be defined.

8 Defining a partial valuation

It is well-known how to give a satisfactory (extensional) semantics for syllogistic logic. Almost all modern textbooks use Venn diagrams, and thus a set-theoretic extensional semantics, to represent the meaning of sentences and to decide whether a syllogistic inference is valid. A sentence like SaP, for instance is counted as true if the individuals in the extension of S are also in the extension of P. Following van Rooij presentation, according to such a semantics $M = <D, E>$ is a model with with D a domain of objects and E an interpretation function

---

[9]([19]: 2), claimed that a syllogism is really a single conditional proposition with a conjunctive antecedent either logically true or not.

which assigns to each primitive term T a non-empty proper subset of D: $\varnothing \neq E_M(T) \subset D$. Moreover, it can be used a broader interpretation function, V_M, which for primitive terms T is like E: $V_M(T) = E_M(T)$. Categorical sentences are counted as true in the expected way: $V_M(SaP) = 1\ iff\ V_M(S) \cap V_M(P) = V_M(S)\ iff\ V_M(S) \subseteq V_M(P)$ and $V_M(SiP) = 1\ iff\ V_M(S) \cap V_M(P) \neq \varnothing$). SoP and SeP are interpreted as the negations of SaP and SiP, respectively. It is said that $\varphi_1, ..., \varphi_n \models \psi$ iff $\forall M$, $V_M(\varphi_1) = 1\ and...and\ V_M(\varphi_n) = 1$, then $V_M(\psi) = 1$. [38] notes that it is well-known that this semantics validates all and only all arguments in classical syllogistic style if and only if they are traditionally counted as valid. Arguably, we have now a valuation on the language of syllogistic reasoning to incorporate in our justification model.

So far so good. But it is weir in the actual setting that we consider as false a problematic proposition, which is rejecting its truth value. Admittedly, as far as we know, it rather lacks a truth valued instead of having a determinate one. Consequently, it is more convenient to base the valuation on an alternative theory that admits the existence of imprecise sets, whose elements are not totally defined. A theory to deal with uncertainty called Rough Sets theory was introduced by Pawlak in 1982. [10] It can be seen as an extension of (standard) set theory, in which a subset of a universe is formalized by a pair of sets, i.e., the lower and upper approximations. These approximations can be described by two operators on subsets of the universe.

We are going to introduce now some definitions and concepts borrowed from [17] in order to apply rough sets to define a (partial) valuation that incorporates the absence of truth value for this problematic propositions. The intuition behind this valuation is that there is an impediment to assign a definite truth-value to this propositions.

Firstly, a most fundamental (and very general) notion of an approximation space must be given. This core notion serves in Tamás Mihálydeák's presentation 'as the set-theoretical background of semantics of partial first-order logic relying on different partial membership functions'. This is the presentation of Mihálydeák that we will quote directly. Firstly, the following definition of a 'partial approximation space' is introduced:

---

[10]See [1].

Definition. The ordered 5-tuple $\langle \mathcal{U}, \mathcal{B}, \mathcal{DB}, l, u \rangle$ is a general partial approximation space with a Pawlakian approximation pair if

1. $\mathcal{U}$ is a nonempty set;
2. $\mathcal{B} \subseteq 2^{\mathcal{U}}$, $\mathcal{B} \neq \emptyset$ and if $B \in \mathcal{B}$, then $B \neq \emptyset$;
3. $\mathcal{DB}$ is an extension of $\mathcal{B}$, and it is given by the following inductive definition:
   (a) $\mathcal{B} \subseteq \mathcal{DB}$;
   (b) $\emptyset \in \mathcal{DB}$;
   (c) if $D_1, D_2 \in \mathcal{DB}$, then $D_1 \cup D_2 \in \mathcal{DB}$;
4. the functions l, u form a Pawlakian approximation pair $\langle l, u \rangle$, i.e.
   (a) $l(S) = \bigcup C^l(S)$, where $C^l(S) = \{B \mid B \in \mathcal{B} \text{ and } B \subseteq S\}$;
   (b) $u(S) = \bigcup C^u(S)$, where $C^u(S) = \{B \mid B \in \mathcal{B} \text{ and } B \cap S \neq \emptyset\}$.

Again, quoting from Mihálydeák's paper, the following definition specify what it is a border set of an imprecise set S.

Definition. If $\langle \mathcal{U}, \mathcal{B}, \mathcal{DB}, l, u \rangle$ is a general partial approximation space with a Pawlakian approximation pair and $S \subseteq \mathcal{U}$, then $b(S) = \bigcup \left(C^u(S) \setminus C^l(S) \right)$ is the border set of S.

This is explained informally as follows: 'the set $\mathcal{U}$ is the universe of approximation; $\mathcal{B}$ is a nonempty set of base sets, it represents our knowledge used in the whole approximation process; $\mathcal{DB}$ (i.e. the set of definable sets) contains not only the base sets, but those which we want to use to approximate any subset of $\mathcal{U}$; the functions l, u (and b) determine the lower and upper approximation (and the border) of any set with the help of representations of our primitive or available concepts/properties. [17].

The nature of an approximation pair depends on how to relate the lower and upper approximations of a set to the set itself. A general partial approximation space can be specified by giving some requirements for the base sets.

What is more useful to our purpose is the following function μ_S^c, the crisp membership function for any set and any object' [17].

Definition. If S ($S \subseteq \mathcal{U}$) and $u \in \mathcal{U}$, then

$$\mu_S^c = \begin{cases} 1 & if\ u \in S \\ 0 & if\ u \in \bigcup \mathcal{B} \setminus S \\ 2 & otherwise \end{cases}$$

The number 2 is used as a null entity to show that a function is undefined for an object u, i.e. to represent partiality of membership functions. Precisely, this is the status that should be assigned to our 'problematic' proposition.

Let us reformulate now the previous assignments of truth values to the atoms of the language. After producing some changes in the function $V_M(T)$ to operate with rough sets now, we have that the predicate 'Negative(x)' is to be represented by a rough set of the domain of objects D. Consequently, it turns out that the truth condition for SeP that corresponds to the negation of SaP depends now on the membership function $\mu_S^c(u)$ just defined.[11] Thus, in the case of NPN we have that μ_S^c must be 2, since for any S it does not belong to S or $\bigcup \mathcal{B} \setminus S$.

Accordingly, the valuation can be changed by admitting a value ⊙ compatible with the outcome corresponding to this null object. Thus, we have to define a new partial valuation $\mathcal{V}_M$. For crisp sets this valuation coincides with the former valuation V_M. But it differs for non crisp or imprecise sets like the set of 'negative propositions' in this scenario. In this case, V_M becomes partial incorporating the value ⊙.

9 Adapting the Basic LJ–Model

In justification logic there are two sorts of logical objects: the so-called justification terms Tm and formulas Fm. Tm has constants and variables. They are also built from constants and variables by an operation '.' called application.

Formulas are built from propositional letters, p, q, r... and boolean connectives. There are also new formation rules: if t is a justification term and φ is a formula, then $t : \varphi$: is a formula (t is a justification for φ).

---

[11] $V_M(\text{SaP}) = 1\ iff\ V_M(S) \cap V_M(P) = V_M(S)\ sii\ V_M(S) \subseteq V_M(P)$.

The logical system J⁻ of basic justification logic (see [2]) consists of a background propositional logic equipped with the rule of modus ponens and the *application*:

$$s : (\varphi \to \psi) \to (t : \varphi \to [s \cdot t] : \psi)$$

To adapt the basic model to operate only with categorical sentences we have to change to the new set of propositional variables. Instead of atomic propositions we have now other kind of atoms (categorical proposition) to form syllogistic-type inferences.

Definition (Basic Model) A basic model, typically ∗, consists of an interpretation of the members of Var, propositional variables, and an interpretation of the members of Tm, meeting conditions given later. Overloading notation, it is used ∗ for a basic model, and also for the interpretations of Var and of Tm in that model. Moreover, the interpretation of a proposition in a basic model is a truth value:[12]

$$\ast : \text{Var} \mapsto \{0, 1\}$$

Keeping the the interpretation in a given M = <D; E> for an interpreted language of primitive terms, we can make ∗ equal to the general interpretation function V_M. As in the basic model for JL, the interpretation of each justification term is a set of formulas. That is,

$$\ast : \text{Tm} \mapsto 2^{\text{Fm}}$$

It is worth to observe that if this language was made self-referential as our initial example, it would be vulnerable to the same paradoxes. However, we will take another direction here.

Definition (Evaluation in a Basic Model) [2] Let be a basic model ∗. This model determines a unique mapping

---

[12] Adapting the model to our present interest, we would have to introduce one more value expressed by ⊙ or as in [4] we can give up the assumption that all sentences receive 0 or 1 and keep the standard two valued assignment.

from all formulas to truth values, $*: \text{Fm} \mapsto \{0,1\}$, meeting the following conditions.[13]

(1) $P* = *(P)\ for\ P \in Var$.
(2) $\perp * = 0$.
(3) $(X \to Y)* = 1$ if and only if $X* = 0\ or\ Y* = 1$ (It could be similar for other connectives).
(4) $(t : X)* = 1$ if and only if $X \in t*$.

The interpretation of a proposition in a basic model is a truth value. That is,[14]

$$*: \text{Var} \mapsto \{0,1\}$$

The interpretation of each justification term is a set of formulas. That is,

$$*: \text{Tm} \mapsto 2^{\text{Fm}}$$

A basic model and its mappings extends to a valuation on all justification formulas.

Models corresponding to J^- can be specified by a combinatorial condition. For set of formulas S and T define:

$$S \triangleright T = \{\varphi \mid \psi \to \varphi \in S\ and\ \psi \in T\ for\ some\ \psi\}$$

Informally, it is the result of applying modus ponens to all members of S and T once in a given order. It follows that this closure condition holds:

$$S^* \triangleright t^* \subseteq [s \cdot t]^*$$

---

[13] Now we would have one more value expressed by ⊚ if it is extended to a partial valuation as, for instance, in [4], where "monotonic functions from $\{T, \star, \perp\}^n$ into $\{T, \star, \perp\}$ can be taken to represent partial functions from $\{T, \perp\}^n$ into $\{T, \perp\}$. Initially, in this semantics "for simple partial logic" as Blamey says, we can adopt classical $T, \perp$ (T-F) conditions; "only we give up the assumption that all sentences have to be classified either as T or as $\perp$. This leaves room for the classification neither-T-nor-$\perp$." A three-valued justification logic is also considered in [20].

[14] As we noted before, now we could also have one more value expressed by ⊚, changing this set to $\{0, 1, ⊚\}$

10 Introducing Justification Models formally

Although our treatment will remain mainly informal, In this section we will sketch some modification to JM, recently worked out by Artemov, in order to adapt them to our specific stage-settings. Instead of models purely elaborated in terms of belief and knowledge, we will change to predicates that produce that the agent 'becomes informed' about something. Specifically, we tray to capture the phenomenon of becoming informed about the signification of the proposition involved in the inference. If justifications s and t bring about the agent becomes informed about something, then their product also brings about it becomes informed about the corresponding formulas.

Thus, we will elaborate on a JM focusing our attention on the predicate of 'becoming informed' related with justifications. We will dub it IP, for 'information producing'. In this way, we are adapting standard JM to our specific purpose.

Definition (A Basic Adapted JM) [2] is a triple $M = (*, A, \text{IP})$ where

1. $*$ is a basic J(CS)-model for some axiomatically appropriate constant specification CS.[15]

2. A is an acceptance predicate, it is accepted as a justification for α, such that
   A is fair to CS: $A(c, \alpha)$ iff $c : \alpha \in \text{CS}$,
   A is closed under application: if A (s, $\varphi \to \psi$) and A (t:φ) then A (s: · t, ψ),
   A is consistent: not A (t, $\bot$) for any t;

3. IP is a information-producing predicate, t is information producing for α, such that, IP is fair to CS: IP (c, α) iff $c : \alpha \in \text{CS}$,
   IP is closed under application: if IP (s, $\varphi \to \psi$) and IP (t,φ) then IP (s · t, ψ).

---

[15] The set of justification constants is the set JCon = {t1, t2,...}. Justification constants are to be thought of as un-analysable primitive pieces of justification that cannot be broken down, but can be recursively combined to form complex justifications. Constants in justification logic are used to denote justifications of assumptions, in particular, axioms.

We can say that a sentence ϕ is believed if it has an accepted justification, i.e., A (t,ϕ) holds for some t. We will not be using this fact though.[16]

To illustrate the solution, we will just consider the first paradox. Let assume that the scenario contains OPA and a version of NPN is implied by the *locus from contraries* (LC), that is licensed to be used in all scenarios. Thus, somehow NPN also appears in the scenario. We have then a proposition OPA that implies another (NPN). That is,

$$OPA \to NPN$$

Whatever be the value of this conditional it won't be false. More to the point, there are also justifications to become informed in the scenario about these implication. Accordingly,

$$(t : (OPA \to NPN))$$

But there is no evidence to support the negation of OPA, namely, $\neg$(OPA). We never become informed in this scenario about the existence of a negative proposition.[17]

Thus, in this scenario the agent accepts a proposition, even though a truth value cannot be assigned to it and becomes informed about something. From the evidence given by OPA and LC_I she accepts NPN and it turns out to be an Information-Producing proposition. The controversial point is how NPN becomes accepted by the agent and convey some information. What is the evidence on which it is based? It turns out that the evidence for the model is based on OPA and LC_I. Informally, in the precise basic model $*$, NPN has no definite value, moreover (s · t) is a justification for NPN; it is accepted for NPN and is information-producing for it regardless of not being true.

---

[16] The notion expressed by means of the predicate 'Information-Producing' can be thought of as the information brought about by the justification attached to the proposition in each case. This justification causes the agent to become informed about this proposition.

[17] What is relevant is the process of justifying NPN that proceeds by using clause 3 supra of the Definition of our basic adapted JM. Let LC_I be the adequate first-order instance of LC. Thus, IP(t, OPA$\to$ NPN) and trivially IP(s, OPA). In consequence, IP(s · t, NPN), with s*={OPA} and t*={OPA$\to$ NPN}. Properly, justifications can be thought in terms of a supposition and LC_I.

Even if a self-referential language was forced as in the former example, there would be no evidence about negative propositions in the actual scenario. There just would be some evidence supporting a proposition, which informs us that there could not be negative proposition according to the scenario. Facts could be as the proposition (the *positum*) says (while God has eliminated all negative propositions). Admissibly, one just becomes informed -through the justification provided by the terms- about a negative proposition which says that. What is the reason to admit that NPN is negative? One can say that it is the reason concerning the information about the negative character of NPN. But, it just inform us that a negative proposition serve to communicate that there are no negative propositions according to the actual stage-setting. This justification cannot afford now a reason to admit the existence of a negative proposition that contradicts the premise.[18]

11 Conclusion

It will be convenient to settle the inferences of Buridan's examples in a concrete stage-setting, since there are many particular features in each case that have to be taken in account. Earlier, we have in turn developed this stage-setting-based approach to connect our first scheme to a more sophisticated treatment, by applying Artemov's stage-settings concerning special justification models for justification logics. That is to say that, at the end of the trip, we have also put the resolution of Buridan's paradoxes in this setting, elaborating on models earlier introduced by Artemov. It is important to understand that Justification Models do not analyze why certain justifications are accepted, but rather they assume accepted justifications to be given and provide a formal framework for reasoning about them. ([2]: 227)

Analogously, it is worth to note that our approach using justification

---

[18] An interesting topic to continue our research is a reformulation of the *Obligatio* rules; in which, based on the above, more complex notions of 'follow' and 'granted' can be thought of. Thus the first three rules proposed by Burley could be reformulated. (B1) The first proposition put forward by the opponent, the *positum*, must be granted. (B2) Everything proposed that follows from already granted propositions must be granted. (B3) Every sentence proposed, whose negation follows from already granted propositions must be denied. [14]

models do not intend to analyse why certain inferences do not preserve truth or why they do preserve signification, 'firmness' or 'reference', but rather this model assume signification-preserving inferences to be given and provide a formal framework to take account of them.

To start with, we assumed that all sentences or propositions are given inside a stage-setting, a stage-setting with truth-valued-impediments wherein sentences that lack a definite truth-value are also allowed. Moreover, it is assumed that sentences have a justification in the stage-setting, which can change from one stage-setting to another. It can serve also to think of an agent to whom justified assertions are attributed inside this stage-setting. In both examples we have appealed to the construction of stage-settings, since more evidence, in the form of self-referring propositions must be added in both cases. Admittedly, the reconstruction is defensible, since the inference is mildly adapted to this background.

In the outcome, Buridan finds some special troubles that confronts the relation of consequence in a language containing self-referring sentences (these are the basic features to consider when analyzing PS). Approaching the problem with the aid of stage-settings designed with an specific purpose and based on Artemov's approach can solve the paradoxes by incorporating the basic inference process. Finally, it is worth to note that as long as Buridan's examples exceeds the frame of purely formal reasoning, some features of material inference have to be incorporated in the process of formalization. Precisely, these aspects are best accommodated by appealing to the introduction of well-designed stage-settings.

Analyzing Buridan's paradoxes we have seen that preservation of truth can fail in a language with self-referring propositions of certain type. We have also sympathized with the thesis that it is most important to preserve not truth, but significations and this is our link with the idea of 'firmnes' or 'correspondence' (Section 4). In these cases, all we can hope is that the conclusion of the inference signifies what is signified in the premises. We have to reason now with what we become informed about. It does not need to be something true, but mere information what is referred to by the premises and the conclusion. We have some evidence or justification for accepting this information and proceed according to this support. Inference rules concerning propositions or sentences can remain as they are but some special restrictions are introduced to manage evidence or justification that condition these inferences. One is not inferring now with the truth of the propositions but with some kind of information about them. As we have said before what should concern us here is the notion of semantic information and the role it plays in the problem that we paid attention to.[19]

A crucial point of our approach is that the sentence which claims that every proposition is affirmative (the *positum*) should not contain information about the existence of negative propositions. It seems hard to admit that this last piece of information is contained in the premise of the inference in spite of what is implied by the conclusion. Granting the logical transition from the premise to the conclusion, transitivity seems to be involved as well. Moreover, if we want to keep intact all the properties of classical inference, we should leave truth aside and think only in terms of information.[20] If it is signified by the premise that every proposition in the actual stage-setting is affirmative, it cannot be consistently signified by this sentence in the very same stage-setting that there is also a negative proposition. We cannot legitimately extract this information, even if a negative proposition were logically implied by the premise. According to Buridan's way of thinking, we have not been told that a negative proposition exists. In other words, we never became informed about that.

Introducing justifications in the stage-setting throughout awareness models tends to make clear what is allowed in the stage-setting according to the information introduced by the sentences. If things are now as they are signified by the premise of the inference in the stage-setting, then not all information is available to the agent therein. Accordingly, some sentences are explicitly justified by the available information, but there could be sentences that are not justified. Admissibly, in Buridan's examples a sentence is logically inferred from other sentences by means

---

[19]In this sense, this work can be seen as an alternative solution to the problem considered by [12] and [7].

[20]The latter makes our proposal consistent not only with Ashworth's comments (see quoted in *section 1*, but with what Buridan himself thought: 'Buridan himself says that, in practice, the final definition of consequence is needed in only a few cases; for most cases, the familiar definition in terms of truth-values is perfectly sufficient' ([7]: 296).

of rules that guarantee truth-preservation. The algorithm that we have mentioned before serve to fulfil this requirement. At the end, this algorithm outputs a sentence that must be true. It is an unassailable truth, but is not allowed according to the information conveyed by the premise of the argument in the actual stage-setting. Admittedly, information does not entail truth in the actual setting. Hence, one could get out of the proving process once the conclusion was reached, as far as it goes astray with the information given by the premise. It can go much further. What is most important now is what information we are reasoning about. If an inference is the process of drawing a conclusion from supporting evidence, then models will serve in our case to control those inferences that can be drawn by an agent in a concrete stage-setting.[21]

References

[1] Akama, S., Kudo, Y., Murai T., *Topics in Rough Set Theory*. Springer. 2020.

[2] Artemov, S. and Fitting M., *Justification Logic*. Cambridge University Press, 2019.

[3] Ashworth, E. J., 'Was Buridan a Psychologist in His Logic?' in Klima (ed.) *Questions on the Soul by John Buridan and Others, A Companion to John Buridan' s Philosophy of Mind*. Springer Int. Publishing. 2018.

[4] Blamey, S., 'Partial logic', in D.M. Gabbay and F. Guenthner (eds.), *Handbook of Philosophical Logic*. Kluwer Academic Publishers. 2002, 2nd Edition, Volume 5, 261-353.

[5] Buridan, J. *Summulae de Dialectica*. Yale University Press, New Haven & London, 2001.

[6] Buridan, J. *Treatise on Consequencies (Tractactus de Consquentiis)*. Transl. by S. Read Fordham University Press, 2015.

[7] Dutilh Novaes, C., 'Buridans Consequentia: Consequence and Inference Within a Token-Based Semantics' in *History and Philosophy of Logic*, 2005, 26, 4, pp. 425–442.

[8] Gabbay, D., Schild U., and David, E., 'The Talmudic Logic Project, Ongoing Since 2008' in *Logica Universalis,*, 2019, 13, pp. 425–442.

---

[21]Work partially supported by Secyt UNC, Proyecto Consolidar 2018-2022 (30720130100358CB).

[9] Haack, S., *Philosophy of logics*. Cambridge University Press, Cambridge, 1978.

[10] Hughes, G. E., *Chapter Eight of Buridan's Sophismata. Translated, with an Introduction and a Philosophical Commentary*. Cambridge University Press, 1982.

[11] Klima, G., 'Consequences of a closed, token-based semantics: the case of John Buridan' in *History and Philosophy of Logic*, 2004, 25(2), 95–110.

[12] Klima, G. , 'Consequence' in C. Dutilh Novaes & S. Read (eds.) *The Cambridge Companion to Medieval Logic*. Cambridge University Press, 2016.

[13] Kulicki, P., 'Aristotle' s Syllogistic as a Deductive System' in *Axioms*, 2020, 9(2).

[14] Lagerlund, H. and Olsson, E., 'Disputation and Change of Belief-Burley's Theory of Obligationes as a Theory of Belief Revision' in H. Lagerlund (Ed.), *Encyclopedia of Medieval Philosophy*. Dordrecht: Springer, 2001, pp. 550-551.

[15] Martin, C., 'Obligations and Liars', in H. Lagerlund (Ed.), *Encyclopedia of Medieval Philosophy*. Springer., Dordrecht, 2001, pp. 550-551.

[16] Maudlin, T. *Truth and Paradox. Solving the Riddles*. Clarendon Press, Oxford. 2006.

[17] Mihálydeák. T., 'Aristotle's Syllogisms in Logical Semantics Relying on Optimistic, Average and Pessimistic Membership Functions' in Chris Cornelis Marzena, Kryszkiewicz, Dominik Slezak, Ernestina Menasalvas Ruiz, Rafael Bello, Lin Shang (Eds.) *Rough Sets and Current Trends in Computing*. Springer, 2014.

[18] Muñoz Delgado, V. L., *La lógica nominalista en la Universidad de Salamanca (1510–1530)*, Publicaciones del Monasterio de Poyo, Madrid, 1964.

[19] Łukasiewicz, J. *Aristotle's Syllogistic from the Standpoint of Modern Formal Logic*. Clarendon Press, Oxford, 1951.

[20] Lurie, J., *New Directions in Justification Logic*. Doctoral Dissertation, University of Connecticut, 2018.

[21] *Paulus Venetus Logica Parva. R. Perreiah (tr.)*. Philosophia Verlag, Munchen-Wien, 1984.

[22] Patzig, G., *Aristotle's Theory of the Syllogism, A Logico-philological study of Book A of the Prior Analytics*. Springer, 1968.

[23] Pironet, F. and Spruyt, J., 'Sophismata' in E. Zalta (ed.), *The Stanford Encyclopedia of Philosophy*. Winter Edition, 2019, URL = <https://plato.stanford.edu/archives/win2019/entries/sophismata/>.

[24] Pironet, F., 'The Relations between Insolubles and Obligations in Medieval Disputations' in H. Lagerlund (Ed.), *Encyclopedia of Medieval Philosophy*

Springer., Dordrecht, 2001, pp. 550-551.

[25] Primiero, G., ' An episte)mic logic for becoming informed' in *Synthese*, 2009, 167.

[26] Prior, A., 'The Possibly-True and the Possible' in *Mind. New Series*, 1969, 78, 312, pp.481–492.

[27] Read S., Formal and Material Consequence' in *Journal of Philosophical Logic*, 1994, 23, pp. 247–265.

[28] Read S. 'Self-Reference and Validity Revisited' in Yrjönsuuri M. (ed) *Medieval Formal Logic*. Dordrecht, Springer, 2001.

[29] Read, S., The Rule of Contradictory Pairs, Insoluble and Validity', in *Vivarium*, 2020, 58, 275–304.

[30] Scott, T., *Sophisms on Meaning and Truth* (tr. *Sophismata*, by J. Buridan). Meredith Publishing Company, New York. 1966.

[31] Seaton, W. K., *An Edition and Translation of the Tractatus de consequentiis by Rudulf Strode, Fourteenth -Century Logician and Friend of Geoffrey Chaucer. PhD thesis.* University of California, Berkeley, 1973

[32] Spade, P., *Lies, Language, and Logic in the Late Middle Ages.* Vivarium Reprints, 1988.

[33] Spade, P., 'Insolubilia' in N. Kretzmann, A. Kenny, J. Pingborg and Stump (eds.) *The Cambridge History of Later Medieval Philosophy From the Aristotle to the disintegration of scholasticism 1100–1600.* Cambridge University Press, New York, 2008.

[34] Spade, P. and Read, S., 'Insolubles' in E. Zalta (ed.), *The Stanford Encyclopedia of Philosophy* FallEdition, 2018.

[35] Uckelman, S. L., 'Prior on an insolubilium of Jean Buridan' in Synthese, 2012, 188(3), 487–498.

[36] van Benthem, B., *Modal Logics for Open Minds.* CSLI Lecture Notes, Stanford University, 2010.

[37] van Benthem, B., 'Logic Games: From Tools to Models of Interaction' in *Proof, Computation and Agency, Logic at the Crossroads.* Springer Dordrecht, 2011.

[38] van Rooij, R., 'The propositional and relational syllogistic' in *Logique et Analyse*, 2012, 55(217), 85–108.

[39] Wilks, I., 'Peter Abelard and his contemporaries' in D. Gabbay & J. Woods (eds.) *Handbook of the History of Logic Volume 2 Mediaeval and Renaissance Logic.* ELSEVIER, 2008.

[40] Yrjönsuuri, M., 'Treatments of the Paradoxes of Self-reference' in Gabbay, D. & Woods, J. (eds.) *Handbook of the History of Logic Volume 2 Mediaeval and Renaissance Logic*, ELSEVIER. 2008.

[41] Yrjönsuuri, M., 'Insolubles' in H. Lagerlund (Ed.), *Encyclopedia of Medieval Philosophy* Springer, Dordrecht, 2011, pp. 550-551.

[42] Zupko, J., 'John Buridan' in E. Zalta (ed.) *The Stanford Encyclopedia of Philosophy* Winter Edition, 2019.

"Men go grey": Robert Kilwardby and the Logic of Natural Contingency

Joshua Mendelsohn
Loyola University Chicago
jmendelsohn@luc.edu

Logical compendia and commentaries on Aristotle's *Prior Analytics* in the twelfth and thirteenth centuries divide the senses of "possible" in a standard way. First, possibility is divided into a broad sense in which only the impossible is excluded (but which is compatible with necessity), and a narrower sense that excludes both necessity and impossibility (i.e., a sense of "possible" meaning "neither necessary nor impossible").[1] Although medieval logicians did not use any notation resembling that of contemporary modal logic, it is not hard to see how to express this distinction using contemporary notation: If we use $\Diamond$ to represent possibility in the broad sense, then a statement p is possible in the narrow sense if and only if $\Diamond p \wedge \Diamond \neg p$. I will refer to the narrower sense of possibility as

---

For their input during the writing process, I am indebted to the members of the 2018 Doctoral Workshop on Medieval Logic in Cologne and the 2021 Medieval Logic and Ontology Workshop at KU Leuven, as well as Wolfgang Lenzen, Simon Babbs, and an anonymous reviewer. I would also like to thank Paul Thom for generously sharing drafts of his work, for raising questions which led to numerous improvements, and for his ongoing mentorship and support.

[1] See 'Anonymus Aurelianensis III' in Aristotelis *Analytica priora* [36, pp. 98.3-6, 99.27-28], Lambert of Auxerre [1, pp. 39-41], Robert Kilwardby [35, pt. 1, p. 376.235-243], Albert the Great [4, tract. I, c. 12]. Some authors in this period also, confusingly, delineate a third sense of possibility that they identify with necessity. This may be the result of confusing the claims that some possibilities in the broad sense are necessary (which is true) with the claim that "necessary" is one meaning of "possible" (which is false): On this see Thom [31, pp. 29-30] and Lagerlund [16, p. 23]. I will set aside this putative sense of possibility in what follows.

"contingency".[2]

Contingency is then further divided into at least two sorts: "Indeterminate" contingency (*contingens infinitum*) and "natural" contingency (*contingens natum*).[3] This distinction tends to be glossed in terms of how the contingency is related towards being and non-being. "Indeterminate" contingencies, which are illustrated by chance events, are said to be related "equally" towards being and non-being,[4] whereas a "natural" contingency is said to be related "more" towards being,[5] or "unequally" towards being and non-being.[6] The stock example of the latter type of contingency was "men go grey in old age", or sometimes just "men go grey".[7]

---

[2] Modern scholars of medieval logic sometimes also call this "two-sided" [31, p. 138] or "two-edged" [14, p. 531] possibility. I follow the usage of Thom [34] in calling the narrow sense "contingent". Most logicians of the thirteenth century do not distinguish the terms *possibile* and *contingens* in this way, since these were Boethius's translations of Aristotle's terms δύνατον and ἐνδεχόμενον respectively, and neither Aristotle nor Boethius employed these terms to mark the distinction between broad and narrow possibility. The sub-types of contingency are, however, usually referred to as *contingens natum* and *contingens infinitum*. On the terminological issues see Knuuttila [15, pp. 106–7], Lagerlund [16, p. 23] and Hintikka [11, pp. 27–40]. I will follow the modern usage and always use "contingent" and "contingency" to refer to narrow possibility, except in quotations.

[3] See 'Anonymus Aurelianensis III' in Aristotelis *Analytica priora* [36, p. 100.2–3], *Dialectica Monacensis* [7, p. 481.14–21], Lambert of Auxerre [1, p. 42.14–26], Robert Kilwardby [35, pt. 1, pp. 368.123–370.127], Albert the Great [4, tract. I, c. 12–15]. *Contingens infinitum* is also sometimes referred to as *contingens ad utrumlibet*, but the latter expression is used in some early texts to refer to possibility in the broad sense rather than indeterminate contingency; on this see Knuuttila [14, p. 532, 15, p. 112], Jacobi [12, pp. 92–4] and Lagerlund [16, pp. 24–25].

[4] *aequaliter se habet ad esse et ad non esse*: Albert the Great [4, tract. I, c. 12], Robert Kilwardby [35, pt. 1, p. 370.136–7]. *non magis se habet ad esse quam ad non esse*: 'Anonymus Aurelianensis III' in Aristotelis *Analytica priora* [36, p. 100.30–31], Lambert of Auxerre, [1, p. 42.21–22]. *sive 'infinitum', sive 'aequale'*: Roger Bacon [6, 2.1, §392]. On choice and chance, see note 11.

[5] *magis se habet ad esse quam ad non esse* [1, p. 42.17].

[6] *non equaliter se habet ad esse et non esse* [35, pt. 1, p. 398.581].

[7] *Hominem canescere in senectute*: Roger Bacon [6, 2.1, §392], 'Anonymus Aurelianensis III' in Aristotelis *Analytica priora* [36, p. 100.10], Lambert of Auxerre [1, p. 42.16]. *Hominem canescere*: Robert Kilwardby [35, pt. 1, p. 370.131] (but see [35, pt. 1, p. 394.530–542], where Kilwardby seems aware of the usual addition *in senectute*).

Unlike the distinction between broad possibility and contingency, the distinction between natural and indeterminate contingency does not readily suggest any translation into contemporary modal logic. Indeed, the sense of the distinction is not really made clear by this standard gloss. In what way does the natural tendency of men to develop grey hair, or to do so in old age, make this a predication which is more "related towards being" than a chance event? What, for that matter, does it mean for a contingent predication to be related "more towards being", and in what way does this render a contingency "natural"?

Robert Kilwardby (c. 1215–1279), a Parisian Master of Arts and later Archbishop of Canterbury, is among the first Latin commentators to go beyond the standard tropes and give an account of why certain contingencies are distinctively natural. He does so in his extensive question-commentary on Aristotle's *Prior Analytics*,[8] in which he also covers the logic of natural contingency, treating it as a modality within a system of syllogistic modal logic on the basis of the meaning he takes the expression *contingens natum* to have.

My goal in what follows is, first, to explain how Kilwardby develops his notion of natural contingency in the course of addressing exegetical problems that arise in interpreting *Prior Analytics* I.13. I then show how Kilwardby's way of understanding the distinction allows him to sketch a logic of natural processes and their results that may be formalized in a temporal free logic. Since Kilwardby gives this account in the course of his commentary on Aristotle, it is important to first review the passage that forms the basis for Kilwardby's understanding of the distinction between kinds of contingency and the exegetical problems that this passage gives rise to.

---

[8]The critical edition and translation of Kilwardby's commentary is due to Thom and Scott [35]. For a brief overview of what is known about Kilwardby's life and career, see Silva [27].

1 Robert Kilwardby's commentary on *Prior Analytics* I.13

In *Prior Analytics* I.13, Aristotle commences his treatment of syllogisms with premises containing a modality of contingency. Aristotle defines "being possible" and "the possible" as "that which, while not being necessary, will not lead to anything impossible when it is assumed to belong" [29, I.13, 32a18–20].[9] This definition rules out both necessity and impossibility, and thus is a definition of what I am calling contingency. After discussing the conversion of contingencies, Aristotle then adds the following remark:

> After these explanations, let us add that 'being possible' is said in two ways: in one way of what happens for the most part, when the necessity has gaps, such as that a man turns grey or grows or ages, or generally what belongs by nature. For this has no continuous necessity because a man does not exist forever, but while a man exists, it happens either of necessity or for the most part. In another way 'being possible' is said of what is indeterminate, that is, what is possible both this way and not this way, such as that an animal walks or that an earthquake happens while it walks, or, generally, what comes about by chance, for this is by nature no more this way than the opposite way [29, I.13, 32b4–13].[10]

This passage announces that it will distinguish two further senses of "being possible", here referring to the narrow sense of contingency. One of these is associated with "what belongs by nature" and the other with

---

[9]λέγω δ' ἐνδέχεσθαι καὶ τὸ ἐνδεχόμενον, οὗ μὴ ὄντος ἀναγκαίου, τεθέντος δ' ὑπάρχειν, οὐδὲν ἔσται διὰ τοῦτ' ἀδύνατον. For the Greek of the *Prior Analytics* and *Posterior Analytics* I rely on the edition of W. D. Ross [26].

[10]Διωρισμένων δὲ τούτων πάλιν λέγωμεν ὅτι τὸ ἐνδέχεσθαι κατὰ δύο λέγεται τρόπους, ἕνα μὲν τὸ ὡς ἐπὶ τὸ πολὺ γίνεσθαι καὶ διαλείπειν τὸ ἀναγκαῖον, οἷον τὸ πολιοῦσθαι ἄνθρωπον ἢ τὸ αὐξάνεσθαι ἢ φθίνειν, ἢ ὅλως τὸ πεφυκὸς ὑπάρχειν (τοῦτο γὰρ οὐ συνεχὲς μὲν ἔχει τὸ ἀναγκαῖον διὰ τὸ μὴ ἀεὶ εἶναι ἄνθρωπον, ὄντος μέντοι ἀνθρώπου ἢ ἐξ ἀνάγκης ἢ ὡς ἐπὶ τὸ πολύ ἐστιν), ἄλλον δὲ τὸ ἀόριστον, ὃ καὶ οὕτως καὶ μὴ οὕτως δυνατόν, οἷον τὸ βαδίζειν ζῷον ἢ βαδίζοντος γενέσθαι σεισμόν, ἢ ὅλως τὸ ἀπὸ τύχης γινόμενον· οὐδὲν γὰρ μᾶλλον οὕτως πέφυκεν ἢ ἐναντίως.

"what is indeterminate", which includes what "comes about by chance".[11] "Men go grey" is given as an example of the first kind of contingency, a contingency in the sense of a natural occurrence, which following medieval usage I will call a natural contingency.

Aristotle's characterization of natural contingency in this passage gives rise to significant interpretative difficulties. In particular, it is unclear whether Aristotle takes natural contingencies to be contingent in the usual sense that entails being neither necessary nor impossible. On first glance, Aristotle seems to be identifying natural contingency with "what happens for the most part" [29, I.13, 32b5–6].[12] In that case, his point would be that "men go grey" describes more than a chance occurrence because *most* men go grey (and so it is certainly not impossible); nevertheless, developing grey hair is not a universal or inevitable fate for men and so "men go grey" also falls short of necessity. The predication would then be contingent in the usual sense which requires being neither necessary nor impossible.

What complicates this reading is that Aristotle goes on immediately to say that in the type of contingency exemplified by men going grey, "the necessity has gaps" [29, I.13, 32b5].[13] This suggests that Aristotle means to treat the fact of a man going grey as a type of necessity (albeit one with "gaps"),[14] and so, if it is also a "contingency" it cannot be a contingecy in the usual sense that excludes necessity. The following sentence at first seems to confirm this: Aristotle denies that men going grey has "continuous" [29, I.13, 32b8][15] necessity, apparently suggesting that it does have some sort of "non-continuous" necessity. Yet Aristotle then seems to retreat to the claim that going grey happens "either of

---

[11] Alexander of Aphrodisias takes the latter class of contingencies to include not only chance events but also the results of free choice [38, 162.31-2]. He is followed in the contemporary literature by Smith [28, pp. 126–7]. Medieval authors tend to employ examples of chance occurrences rather than freely chosen occurrences as the contrast with natural occurrences, but Kilwardby does once mention choice (*propositio*) in this connection [35, pt. 1, p. 370.138].
[12] ὡς ἐπὶ τὸ πολὺ γίνεσθαι.
[13] διαλείπειν τὸ ἀναγκαῖον.
[14] Cf. Striker [30, pp. 157–159], discussed below.
[15] συνεχές.

necessity or for the most part" [29, I.13, 32b10–11].[16]

This is rather confusing, and can give the impression that Aristotle hedges or equivocates regarding the modal status of "men go grey" and the other examples of natural occurrences given here. The unclarity recurs a few lines below, where Aristotle asks whether there can be scientific knowledge and demonstration of contingencies. The question arises naturally given his view in the *Posterior Analytics* that scientific knowledge and demonstration concerns only what holds of necessity and excludes what takes place on account of chance.[17] In *Prior Analytics* I.13, he maintains that, although there is "no knowledge or demonstrative syllogism of indeterminate things",[18]

> there is knowledge of things that happen by nature, and by and large arguments and investigations are concerned with what is possible in this way [29, I.13, 32b20].[19]

Here, again, Aristotle's answer gives the impression of hedging: What does he mean when he says that "by and large" (σχεδὸν) there is scientific argumentation concerned with what is contingent in the sense of a natural contingency? Is there, or isn't there?

Given these interpretive difficulties, it is not surprising that this passage has elicited a range of interpretations. In contemporary scholarship, this passage has been subjected to close reading by Gisela Striker, who takes Aristotle here to be attempting to reduce natural occurrences to "gappy" necessities, that is, necessities which hold only when some further condition S is satisfied. Aristotle's attempt fails, however, on Striker's view, since the true logical form of such a statement is a conditional under a necessity operator (viz., $\Box(S \to p)$), and Aristotle's syllogistic logic does not give him the resources to properly express this. Aristotle thus, on her view, wavers between treating natural occurrence as a form of contingency and as form of necessity, aware of some of the problems

---

[16] ἢ ἐξ ἀνάγκης ἢ ὡς ἐπὶ τὸ πολύ ἐστιν.

[17] See *Posterior Analytics* I.4, 73a21–23; I.6 74b5–6; I.30 87b19; I.33 88b30–32 [26].

[18] I.e., of contingencies that are *not* of the class exemplified by "men go grey".

[19] τῶν δὲ πεφυκότων ἔστι, καὶ σχεδὸν οἱ λόγοι καὶ αἱ σκέψεις γίνονται περὶ τῶν οὕτως ἐνδεχομένων

involved in each of these approaches but unable to overcome them [30].[20]

Latin commentators in the twelfth and thirteenth centuries, by contrast, seldom entertain the possibility that Aristotle's view is incoherent, and strive to find a consistent interpretation of the text as they receive it.[21] Among the earliest Latin commentators to attempt an exegesis of this passage are the author of the 'Anonymus Aurelianensis III' and Robert Kilwardby.[22] They read this passage in Boethius's translation. I reproduce the Florence recension of Boethius's translation of 32b4–13 here:[23]

> Determinatis autem his rursum dicimus quoniam 'contingere' duobus modis dicitur, uno quidem quod plerumque fit et deficit necessarium, ut 'canescere hominem' vel 'augeri' vel 'minui', vel omnino 'quod natum est esse' (hoc enim non continuum quidem habet necessarium eo quod non semper est homo, cum autem homo est aut ex necessitate aut ut in pluribus est); alio autem infinitum, quod et sic et non sic

---

[20] Another way in which Striker's interpretation differs from the that of the medieval commentators discussed here is that they all take Aristotle to be distinguishing senses of the term "possible", whereas on Striker's view, Aristotle may be instead identifying the types of case in which a contingency holds [30, pp. 150–151].

[21] This is not to say that commentators in this period take Aristotle to be infallible. Kilwardby, despite his generally reverent attitude towards Aristotle, is occasionally willing to correct him: See Thom [33, p. 256].

[22] On the dating of the former see note 27 below. On the dating of the latter, see Lagerlund [16, pp. 19–21], according to which Kilwardby's commentary predates Albert the Great's, and probably also Roger Bacon's and Lambert of Auxerre's. The texts edited by de Rijk [7] contain discussion of natural contingency but no systematic interpretation of *Prior Analytics* I.13.

[23] This is probably not, word for word, the text that these commentators read. According to Thom and Scott [35, pp. lxxiv–lxxvi], the collated manuscripts of Kilwardby's text also contain readings that agree with the Chartres recension, and sometimes they deviate from Boethius's translation altogether. Thörnqvist argues that the author of the 'Anonymus Aurelianensis III' worked from a conflated version, containing readings from the Chartres and the Florentine editions of Boethius [37, p. 29]. At least as concerns this passage, I have not found any evidence that these commentators were reading a text that differed from the text that I reproduce here in ways that would affect its sense. Here I am primarily interested in the way the text they read renders διαλείπειν τὸ ἀναγκαῖον, and at least on this point both seem to read a text which agrees literally with the Florentine recension.

> possibile, ut 'animal ambulare' vel 'ambulante fieri terrae motum' vel omnino quod a casu fit; nihil enim magis sic natum est vel e contrario.[24]

Boethius translates the phrase διαλείπειν τὸ ἀναγκαῖον, which Striker renders with "the necessity has gaps", as *deficit necessarium* ("lacks necessity" or "falls short of necessity").[25] This has an important effect on the reception of this passage among those who read it in Boethius's translation. Rather than suggesting that "men go grey" is a queer kind of necessity (one with "gaps", as Striker proposes [29, p. 131])[26] Boethius's text says rather unambiguously that it is *not* necessary. Readers who receive *Prior Analytics* I.13 in this translation therefore have an important interpretive parameter set for them: They must make sense of Aristotle's remarks, including Aristotle's invocation of necessity in the following lines, under the assumption that natural contingencies are *not* necessary.

It is therefore not surprising that the author of the 'Anonymus Aurelianensis III', the earliest known Latin commentary on the *Prior Analytics*,[27] takes for granted that the type of contingency exemplified here by "men go grey", "men grow" and "men shrink" are not necessities. He adds, however, that they are in fact also not really contingencies.[28] Instead, he claims, a so-called "natural" contingency is in fact "neither truly

---

[24] *Analytica Priora*, trans. Boethius, recensio Florentina [22, p. 26]. Both commentators read *deficit necessarium* with the Florentine rescension, whereas the Chartres rescension has *diminutum a necessario* [21, p. 160].

[25] Smith [28, p. 18] also translates διαλείπειν τὸ ἀναγκαῖον as "falls short of necessity" in his English translation, but it is not clear that διαλείπειν can have this meaning in Greek: See the entry in Liddell, Scott and Jones [17] and the extended case for a different translation in Striker [30].

[26] On Striker's interpretation, Aristotle's point is that, although it is not true that all men turn grey of necessity, "one might still say that they do age or turn grey of necessity if they live long enough" [29, p. 131]. Hence there is a necessity involved in the greying of men, albeit one with "gaps". We will see that Kilwardby's interpretation also makes use of the idea of an interruption to process that otherwise necessarily produces a certain outcome, but without assimilating natural contingency to necessity.

[27] Thörnqvist [36, p. 2] estimates a date of composition between 1160 and 1180. The commentary is probably based on a lost Greek model [8].

[28] *Sed nota contingens naturale non re uera esse contingens, sed sic dici, eo quod, cum neque sit necessarium neque contingens neque impossibile, aliquo tamen nomine oportuit ipsum appellari et propter affinitatem quidem, quam habet cum contingenti, appellatur contingens* [36, p. 101.7–10].

necessary nor truly contingent, but in a certain way intermediate".[29] That is, "natural contingency" is not a true type of contingency, but rather an intermediate modal status that Aristotle wishes to distinguish from *both* necessity and from contingency proper.

In the twentieth century, Albrecht Becker also defended the view that this passage is intended to distinguish "natural" contingencies from contingencies proper. This reading is however difficult to reconcile with the text.[30] These comments are, after all, explicitly framed as a discussion of the ways that "being possible" is said [29, I.13, 32b4–5],[31] where the context demands that we understand "possible" in the narrow sense of contingency. Aristotle does not deny, as we would expect him to on this reading, that natural contingencies are *contingent*; he rather says that they "have no continuous *necessity*" [29, I.13, 32b8].[32]

In any case, the 'Anonymus Aurelianensis III' does not tell us much about what this supposedly intermediate modal status consists in. The author does make the intriguing remark that indeterminate contingencies are said to be "indeterminate" because "they do not have a natural determinate end to which they are more related".[33] This suggests a teleological understanding of natural contingencies as descriptions of goal-directed occurrences. It also suggests that this is the sense in which they are related more "towards being". The anonymous commentator does not however develop these ideas further.

Robert Kilwardby also takes for granted that natural contingencies are not necessities. Unlike the author of the 'Anonymus Aurelianensis III', however, he emphasizes that natural contingencies are genuine

---

[29] *nec uere est contingens nec uere est necessarium, sed quodam modo medium* [36, p. 100.3–4].

[30] Becker admits that on his reading the discussion is out of place and conjectures that 32b18–22 may be a later insertion [3, p. 77].

[31] τὸ ἐνδέχεσθαι κατὰ δύο λέγεται τρόπους. Boethius: *'contingere' duobus modis dicitur* [22, p. 27]. Cf. Striker [30, pp. 151–152n1].

[32] οὐ συνεχὲς μὲν ἔχει τὸ ἀναγκαῖον. Boethius: *non continuum quidem habet necessarium* [22, p. 27].

[33] *non habet secundum naturam certum finem, ad quem magis se habeat* [36, pp. 100.31–101.1].

contingencies, no less so than indeterminate contingencies.[34] Natural contingencies are, however, a special type of contingency for Kilwardby because the predication they express has a "natural cause".[35] What Kilwardby seems to mean is that, in a natural contingency, there is a causal link between the nature of the subject and predicate (for instance, between the nature of men and developing grey hair). This causal link is what privileges the occurrence of this contingent event over its non-occurrence.[36] It is in other words natural for humans to develop grey hair because being human is the ultimate cause of grey hair.[37]

For Kilwardby, this raises the question why a natural contingency is not simply a necessity, given that the nature or essence of the subject causes it to have the property described in the predicate term. The question is especially pressing for Kilwardby, since he elsewhere defends an essentialist account of syllogistic necessity as reducing to one of Aristotle's first two modes of essential or "*per se*" predication.[38]

His way of dealing with this problem is to distinguish two readings of the predicate in a statement like "men go grey" and the modal statuses of the disambiguated predications. On the one hand, the predicate may be interpreted to denote the *process* of going grey.[39] On this reading, Kilwardby holds, the predication is not a contingency of any sort but rather a necessity since it is necessarily true that men are always in the

---

[34] *tam contingens natum quam infinitum sit non necessarium et possit esse et non esse* [35, pt. 1, p. 396.545–546].

[35] *causam naturalem* [35, pt. 1, p. 370.129].

[36] Kilwardby says that natural contingencies require a "cause dedicated more to one side [*causam ordinatam magis ad unam partem*, i.e., to the predication holding rather than not holding]" [35, pt. 1, p. 396.565–566].

[37] *Dicit igitur quod contingens prediffinitum quoddam est natum, scilicet quod habet causam naturalem ordinatam ad ipsum* [35, pt. 1, p. 370.128–131].

[38] "For necessary propositions reduce to some mode of *per se* inherence, following Aristotle's statement in *Posterior Analytics* I that 'Only *per se* inherences are necessary'." *Propositiones enim necessarie reducuntur ad aliquem modum inherendi per se, secundum quod dicit Aristoteles in primo Posteriorum, 'Sola per se inherentia sunt necessaria'.* [35, pt. 1, p. 160.458–461]. See Mendelsohn [20], Thom [33, pp. 108–125] and Thom [31, pp. 19–25].

[39] *motum in canitem* [35, pt. 1, p. 394.535].

process of going grey.[40] Since this is a necessity in a perfectly strict sense, it may even feature as a theorem of a demonstrative science. Kilwardby offers a chain of causes which suggest he would take the relevant demonstration to be as follows:[41]

1. Greying holds of the incorporation of phlegm into the upper part of the head,
2. the incorporation of phlegm into the upper part of the head holds of the loss of natural heat,
3. the loss of natural heat holds of all men.
4. Therefore, greying holds of all men.

This demonstration explains why men are necessarily in the process of going grey by recourse to their necessary loss of heat. This loss of natural heat takes the form of an incorporation of phlegm into the upper part of the head, which as such causes hair to become grey. The premises of the demonstration are meant to be *per se* predications, that is, necessary truths that hold on account of the essences of men and the various processes described (greying, losing natural heat, etc.). While it is strictly necessary for each of these *processes* to occur in men, it is not necessary for them to come to completion in men. Their completion is merely a frequent occurrence,[42] since the process of greying in men can be "obstructed".[43] The nature of men therefore does not *invariably* bring about the presence of grey hair [35, pt. 1, p. 370.131–135]; this only takes place given sufficient time for a man's hair colour to naturally fade, and a man might die before this occurs.[44]

---

[40] *Si dicat motum in canitiem sic semper ex necessitate canescit homo cum est* [35, pt. 1, p. 394.535–536].

[41] *Prouenit enim canities ex incorporatione fleumatis in superiori parte capitis, cuius incorporationis causa est diminutio caloris naturalis, et ista incorporatio et caloris diminutio semper fit et continue.* [35, pt. 1, p. 394.537–540].

[42] *Si autem dicat terminum motus completum sic ut frequenter canescit homo.* [35, pt. 1, p. 394.540–541].

[43] *impediri* [35, pt. 1, p. 370.130].

[44] *causam ordinatam naturalem sed potest impediri quia non semper est homo* [35, pt. 1, p. 380.132–133]. The context indicates that *non semper est homo* should be taken here to mean that "a man does not exist eternally" rather than "there are not always any men", as Thom and Scott [35, pt. 1, p. 385] translate it. The relevant *dubium* is about whether the statement that a man goes grey requires a man to exist

For this reason, in order for "men go grey" to state a necessity, "grey" must be understood to denote the process of going grey, not the completion of this process.[45] If "grey" is understood to stand for the completion of the process of getting grey hair, then the statement "men go grey" is *not* necessary but only contingent [35, pt. 1, p. 394.540–541]. It is, however, a special type of contingency owing to its connection with this necessary process. Like a necessity, the contingent fact that men go grey derives ultimately from the natures or essences of men and going grey. It however "falls short of a necessity",[46] as Aristotle says on Boethius's translation, because the predicate is not guaranteed to hold of the subject by the causal link that exists between predicate and subject.[47]

Natural contingencies are thus, on Kilwardby's interpretation, predications where the predicate describes the completion state of a process that necessarily occurs in its subject but does not necessarily come to completion in its subject because the subject may not exist for long enough for this process to reach its natural culmination.[48] While in no way necessary, it follows on this analysis that every natural contingency is closely related to a necessity that may often even be expressed using the same terms. That is, while *actually becoming grey* is predicated only contingently of men, this contingency is natural because *undergoing*

---

when he becomes grey. The doubt does not require entertaining the more remote possibility that all men should at some point cease to exist.

[45] *terminum motus* [35, pt. 1, p. 394.535].

[46] *deficit a necessario* [35, pt. 1, p. 370.133–134].

[47] See further Thom [31, pp. 32–34] and Thom [34, pp. 150–152], which my reading so far stands in agreement with. Thom does not emphasize the distinction between process and completion readings of the predicate term as much as I think this should be emphasized, but my principal complaints about Thom pertain to his *formalization* of natural contingencies (discussed below), not his interpretation of what Kilwardby means by *contingens natum*.

[48] It is not clear that Aristotle takes the perishing of the subject to be the *only* way that a natural process can be obstructed. Striker [30, p. 158] proposes going bald as another way the process could be obstructed. It is hard to reconcile this with Kilwardby's view that the process of greying takes place of necessity as long as the subject is alive. Kilwardby in any case assumes that death is the central type of obstruction Aristotle has in mind [35, pt. 1, p. 380.132–133], and I will work with the assumption that all obstructions of natural processes take the form of their subject perishing in what follows.

the process which terminates in grey hair holds necessarily of men.[49] It is Kilwardby's view that either the process or the completion may be understood by the predicate "goes grey"; consequently, either a necessity or a contingency may be understood by "men go grey", depending on the reading given to the predicate term.

This account allows Kilwardby to avoid collapsing natural contingency into occurrence for the most part.[50] He may be contrasted in this respect with his contemporary Roger Bacon, who introduces "contingent in most cases"[51] and "naturally contingent"[52] as synonyms. Bacon draws a threefold distinction among contingencies between those which are "equally" related to being and non-being (like you sitting), those which occur in few cases (like discovering treasure while digging) and those which occur in most cases or "naturally" (like men going grey).[53] Bacon's scheme suggests a statistical, or proto-probabilistic understanding of natural contingency as a contingency which occurs frequently or with high probability.[54]

Kilwardby, by contrast, claims that there are occurrences which are

---

[49] It is useful to compare Kilwardby's notion of natural contingency with what Freddoso [9, p. 225] calls a "deterministic natural tendency", at least as long as we bracket Freddoso's requirement that "only *free* causes can prevent deterministic natural tendencies from blossoming into full-blown natural necessities" (emphasis in original). Kilwardby does not seem to recognize any requirement that the impediment to a natural contingency always be a result of free agency, but he otherwise seems to conceive of the relationship between this type of modality and necessity in a way that is very close to Freddoso's analysis. I am grateful to Elena Baltuta for drawing my attention to Freddoso's work.

[50] Cf. Thom [31, p. 32] contra Lagerlund [16, p. 45].

[51] *contingens ut in pluribus* [6, 2.1, §392].

[52] *contingens natum* [6, 2.1, §392].

[53] *Et illud contingens potest esse tripliciter: aut enim aequaliter se habet ad esse et ad non-esse, et tunc vocatur 'contingens ad utrumlibet' sive 'infinitum' sive 'aequale', ut 'Te sedere'; aut accidit in maiori parte, et tunc vocatur 'contingens ut in pluribus' sive 'contingens natum', ut 'Hominem canescere in senectute'; aut accidit in minori parte, et tunc dicitur contingens ut in paucioribus, ut 'Fodientes invenire thesaurum'* [6, 2.1, §392]. Note that Roger Bacon does not give this division in the course of commenting on *Prior Analytics* I.13. According to McCall [19, p. 68], a similar threefold distinction is found in Averroes.

[54] See Knuuttila [15, pp. 99–137] and Jacobi [13, pp. 18–20] for the statistical and proto-probabilistic interpretations respectively.

natural in most cases (like men going grey in old age) as well as those which are natural in few cases (his example is men going grey in young age).[55] On Kilwardby's view, it is natural for men to go grey in old age because there is a necessary process in men that culminates in greyness. The fact that becoming grey in *old age* is natural in most cases merely marks the fact that this process typically takes a long time (in comparison with a man's lifetime) to complete. It is possible, if uncommon, for the process to come to completion early (and in that case greyness takes place in a young man) [35, pt. 1, p. 398.586–7] as well as for it not to come to completion within a man's lifetime at all (in which case it fails to take place in that man at all) [35, pt. 1, p. 370.132–4]. The relative rarity of these outcomes does not however render them "unnatural": Kilwardby allows that in those cases where a man does go grey in youth, this is likewise a natural occurrence.[56] He thus preserves a distinction between the merely statistical notion of occurrence in most cases or "for the most part", and a separate, orthogonal notion of a natural occurrence.[57]

These interpretive manoeuvres allow Kilwardby to make sense of Aristotle's remarks concerning natural contingency as he receives them. Kilwardby's denial that any contingencies, natural or not, serve as scientific premises, agrees with Aristotle's claim in the *Posterior Analytics* that the premises of demonstrations are necessities.[58] At the same time, he is able to make sense of Aristotle's statement in *Prior Analytics* I.13 that "by and large arguments and investigations are concerned with what is possible in this way" (i.e., naturally contingent) [29, I.13, 32b20–21].[59] These words come to Kilwardby in Boethius's translation as *et paene orationes et considerationes fiunt de sic contingentibus* [35, pt. 1, p. 372.178–9]. Kilwardby understands *paene* here in the sense of "roughly

---

[55] Cf. Thom [34, p. 150].

[56] *canescere in iuuentute quod est natum in paucioribus* [35, pt. 1, p. 398.586–7].

[57] For this reason the statistical interpretation of natural contingency presented in Knuuttila [15, pp. 99–137] and Jacobi [13, pp. 18–20] does not apply to Kilwardby. Thom [31, p. 32] rightly objects to the attribution of this view to Kilwardby in Lagerlund [16, p. 45].

[58] *Posterior Analytics* I.4, 73a21–24. For Kilwardby's commentary on this chapter, see Cannone [5].

[59] σχεδὸν οἱ λόγοι καὶ αἱ σκέψεις γίνονται περὶ τῶν οὕτως ἐνδεχομένων

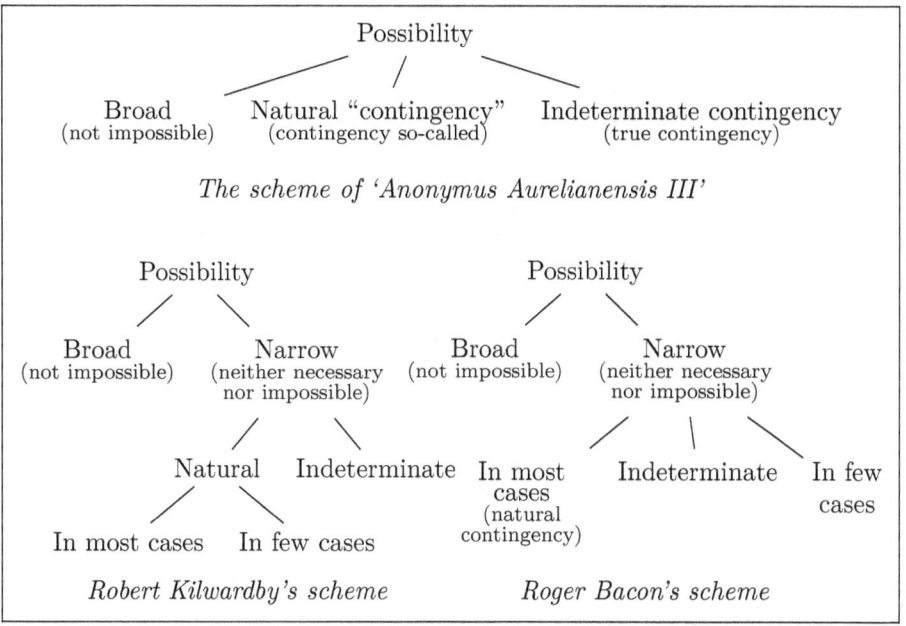

Figure 1: Divisions of possibility

speaking".[60] According to Kilwardby, Aristotle employs the qualification "roughly speaking" because "there are demonstrations about such things, not insofar as it is possible for them not to be so, but insofar as they are necessary" [35, pt. 1, p. 372.165–167]. What Kilwardby seems to mean is that, although Aristotle does not in fact hold that there are demonstrations of natural contingencies, he does hold that there are demonstrations of closely related, necessary propositions, namely of the predications describing the processes underlying natural contingencies. Such necessities may, as in the case of "men go grey", even be expressed using the same terms as a natural contingency. We can thus understand why Aristotle might have said that there is "roughly speaking" a demonstration containing the natural contingency: There is, at any rate, a demonstration of a closely related statement (in this case, that men undergo the process of greying), and the syllogistic argument for a natural contingency will employ the same terms as this demonstration,

---

[60]This is how Thom and Scott [35, pt. 1, p. 373], correctly in my view, render the term in Kilwardby's commentary.

albeit with their completion readings intended.

Kilwardby's distinction between process and completion readings of predicates also allows him to make sense of Aristotle's remark that a natural contingency "has no continuous necessity because a man does not exist forever, but while a man exists, it happens either of necessity or for the most part" [29, I.13, 32b8–10].[61] On Kilwardby's reading, the reason that it is merely contingent for men to go grey, rather than necessary, is because the process of greying is not "continuous": It may be "interrupted", in particular when a man dies.[62] Thus, on Kilwardby's reading, Aristotle is saying that, while it is indeed necessary for a man to be *going grey* so long as that man exists, the necessity is not "continuous" for the reason that each man only has a finite lifetime and thus the greying process may not finish within the man's lifetime. As for Aristotle's next remark, that greying happens "either of necessity or for the most part" during a man's life, Kilwardby takes this to be an allusion to the distinction between process and completion readings of natural predicates [35, pt. 1, p. 394.530–542]. If "go grey" is understood to mean "is in the process of going grey", then it is necessary for men to be going grey (notwithstanding the fact that this necessity is "non-continuous" owing to the finite lifespan of men). On the other hand if "go grey" is understood to mean "reaches the stage of actually having grey hair", then this is true merely "for the most part" [35, pt. 1, p. 394.540–541].

Kilwardby is thus able to give answers to the most significant interpretive difficulties this passage presents. It does not follow that Kilwardby's interpretation is superior to modern interpretations of Aristotle that find no answers to some of these questions in Aristotle's text, since it of course remains a possibility that Aristotle's own theory of natural contingency was not fully developed. The interpretive advantage of coherence must be weighed against the interpretive cost of attributing to Aristotle distinctions which are not to be found in the text. In

---

[61] τοῦτο γὰρ οὐ συνεχὲς μὲν ἔχει τὸ ἀναγκαῖον διὰ τὸ μὴ ἀεὶ εἶναι ἄνθρωπον, ὄντος μέντοι ἀνθρώπου ἢ ἐξ ἀνάγκης ἢ ὡς ἐπὶ τὸ πολύ ἐστιν. Boethius: *hoc enim non continuum quidem habet necessarium eo quod non semper est homo, cum autem homo est aut ex necessitate aut ut in pluribus est* [22, p. 27].

[62] *hominem canescere non habere continuam necessitatem eo quod non semper est homo* [35, pt. 1, p. 384.381–2].

particular, as we have seen, Kilwardby's reading turns on a distinction between process and completion readings of terms, and for this distinction Kilwardby does not refer us, as he often does, to any other text in the Aristotelian corpus. Further, as we have seen, Kilwardby reads Aristotle in Boethius's translation, and it is not clear that Boethius's translation is a faithful rendering of Aristotle's Greek here. For these reasons, the notion of natural contingency Kilwardby develops in his commentary on *Prior Analytics* I.13, 32b4–22 is still best viewed as a development of Aristotle's ideas rather than simply an exposition of them; nevertheless, it is a testament to Kilwardby's prowess as an interpreter that he develops a view of natural contingency that is consistent with Aristotle's exact words as he receives them.

2 Kilwardby on the logic of natural contingency

Kilwardby employs the interpretation he develops of the meaning of natural contingency and the conditions under which natural contingencies hold as the basis for his remarks about their logic. His remarks can be divided into two classes: Those about the conversion rules for natural contingencies and those which concern the syllogistic results involving natural contingency as a special modality.

2.1 Conversion

Aristotle does not say much about the logic of natural contingency. The only claim he makes about the logic of natural contingency comes in the course of introducing the distinction between natural and indeterminate contingency, where Aristotle claims that both natural and indeterminate contingencies "convert with respect to opposite premisses" [29, I.13, 32b13–14].[63] This is the language Aristotle uses to describe the rule of "complementary conversion"[64] above: That is, the rule that if it is contingent that all As are B, then it is contingent that no As are B, and if it is contingent that some As are B, then it is contingent that not all As are B [29, I.13, 32a31–32].

---

[63] ἀντιστρέφει μὲν οὖν καὶ κατὰ τὰς ἀντικειμένας προτάσεις.
[64] For this terminology see Ross [26, p. 45].

These rules are reasonable if contingency is understood in the broad sense of neither necessary nor impossible. Yet, as contemporary scholars have pointed out, this rule is invalid if "contingent" means "for the most part" or means "occurs by nature": From the fact that men for the most part go grey it does not follow that, for the most part, men do not go grey; and from the fact that it is natural for men to go grey, it does not follow that it is natural for them not to do so [see 30, p. 149, 2, p. 351].

Kilwardby, like other commentators around his time,[65] recognizes the problem with taking natural contingency to obey this rule. As he says, natural contingencies "have a cause that is dedicated to one side rather than the other — on account of which, the negation does not have such a cause".[66] If the essence of man entails that men are necessarily in the process of greying, then, assuming the essence of man is consistent, it cannot also entail that men are not in the process of greying [35, pt. 1, p. 396.565-7]. Hence, the fact that it is naturally contingent for all men to go grey cannot imply that it is naturally contingent for no men to do so. Consequently the law of complementary conversion is invalid for natural contingency: $AQ_{nat}aB$ does not imply $AQ_{nat}eB$, and $AQ_{nat}iB$ does not imply $AQ_{nat}oB$.[67]

Kilwardby reconciles this observation with the text by denying that

---

[65] See Albert the Great [4, tract. I, c. 12–14] and, especially, the extended treatment of conversion in Lambert of Auxerre [1, pp. 41–49]. Conversion is also a focus in the 'Anonymus Aurelianensis III' [36, pp. 97–101].

[66] *habet causam ordinatam magis ad unam partem, propter quod eius negatio non habet causam talem* [35, pt. 1, p. 396.565-7].

[67] Here and throughout I adopt the notation for syllogistic premises and conclusions used by Thom [34], adapted from the notation that has become standard in studies of ancient and medieval syllogistics since McCall [19]. Modalities are signified by L (necessity), M (possibility), Q (contingency) and X (assertoric); I introduce Q_{nat} (natural contingency) and X_{simp} (temporally unrestricted assertoric). Quantities and qualities of propositions are signified by 'a' (universal affirmative), 'e' (universal negative), 'i' (particular affirmative) and 'o' (particular negative). In writing syllogistic premises and conclusions using this notation, I place the predicate first. Thus "$ALeB$" is to be read as "A holds necessarily of no B", etc. I also use the standard medieval mnemonic names for syllogisms [on which see 24] followed by a string of modalities in order to denote the modality of their premises and conclusion in a modal syllogism. Thus, "Barbara LXL" refers to a Barbara syllogism with a necessity major premise and an assertoric minor, i.e. the argument $ALaB, BaC \vDash ALaC$.

Aristotle meant to state that natural contingencies convert with natural contingencies. Instead, he understands Aristotle's remark that natural contingencies "convert with respect to opposite premisses, but not in the same way" to mean that a natural contingency converts to a possibility in the broad sense [35, pt. 1, pp. 370.149–157, 396.562–3]: If it is naturally contingent for all men to go grey, then it is possible (in the broad sense of not impossible) for none to go grey. In other words, if we write 'M' for possibility in the broad sense, then Kilwardby accepts the rule that $Q_{nat}a$ converts to Me (and $Q_{nat}i$ with Mo).[68] Kilwardby notes that this fits with his view of natural contingency as having a cause that may be *impeded*: If it is possible, say, for all men to become musical, and this expresses a natural contingency, then the converted proposition that it is possible for no men to become musicians is true only because the natural process of musical education can be impeded, not because it is natural for them to fail to do so [35, pt. 1, pp. 154.379–156.392].[69]

---

[68] See note 67 for a fuller description of the notation used here and throughout.

[69] Paul Thom has pointed out to me that assimilating the case of learning music to the same class of contingencies as men going grey risks trivializing the notion of a natural contingency, since this example then requires us to understand the class of things which may "impede" becoming a musician very broadly, as including anything except the class of events where a man seeks to become a musician and undergoes the necessary training. The worry is then that we might as well say that any contingent predication is natural (it's naturally contingent, e.g., for men to become astronauts since there are specific circumstances under which they do so). One possible solution is to take Kilwardby to recognize two classes of natural contingencies, one exemplified by "men go grey in old age" and another by "men become musicians". In that case, my analysis in this paper should be taken to cover only natural contingencies of the first kind. As against this reading, however, Kilwardby gives no indication that he intends only a partial analysis of natural contingencies when he focuses on the example of "men go grey in old age" in his commentary on *Pr. An.* I.13. Further, Kilwardby does say that humans have a "natural power" to become musical [35, pt. 1 p. 154.385], and that there is a "dedicated cause" for the predicate musical to hold of human [35, pt. 1 p. 156.388]. This suggests that Kilwardby might take musical skill to be the normal actualization of a human ability, a skill which humans will inevitably develop given the desire and opportunity to do so. The same could not be said of, e.g., discovering treasure while digging or becoming pale upon illness. In that case, the class of natural contingencies will turn out to be broader than we might first have expected, but not trivially broad. Whether becoming an astronaut will count as a natural contingency will depend on whether we take humans to have natural abilities that normally result in becoming astronauts under the appropriate circumstances.

2.2 Syllogisms

Unlike Aristotle, who leaves aside the notion of natural contingency after introducing it in the passage discussed above,[70] Kilwardby makes judgements about the role that natural contingencies play in syllogistic modal logic. He does not however systematically present an account of their logic. Following the order of Aristotle's exposition, he makes scattered judgements about the cases in which the contingency premises of a syllogism need to be interpreted as natural rather than indeterminate or generic contingencies. Despite lacking any formal exposition of the logic of natural contingency, however, these judgements can be seen to cohere with his theory of the meaning of "contingent" in this sense.

The first set of judgements Kilwardby makes about the syllogistic logic of natural contingencies concerns syllogisms in the first figure with premises of mixed modality. Kilwardby denies that the contingency needs to be understood as a natural contingency in first figure syllogisms with an assertoric major and a contingency minor [35, pt. 1, pp. 458.459–464.542 (*dub.* 6)]. Instead, the contingency premise is to be understood here in the broad sense that includes both indeterminate and natural two-sided contingencies. In other words, Kilwardby claims that Barbara XQM and Celarent XQM are both valid without any special restriction to natural contingencies [35, pt. 1, p. 462.512–514].[71] On the other hand, he maintains that in Celarent and Ferio with a necessity major and a contingency minor, the contingency must be understood as a natural contingency [35, pt. 1, pp. 500.133–137, 516.419–422, 526.561–528.593 (*dub.* 10)].

He takes the contingency in these syllogisms to be a natural contingency in order to block specious counter-examples to these moods. In first-figure syllogisms with a necessity major and an assertoric minor, Kilwardby had argued that the minor premise must be true without temporal restriction (*simpliciter*), that is, it must hold of all times and not merely "as of now" (*ut nunc*) [35, pt. 1, pp. 316.149–318.163]. This distinction recalls to Aristotle's remark in *Prior Analytics* I.15 that the universal assertoric premise of Barbara XQM must be taken "not with a

---

[70]See Becker [3, p. 77]; Striker [30, p. 150].
[71]See further Thom [33, pp. 85–88]. On the notation used here, see note 67.

limitation of time such as 'now' or 'at such-and-such a time', but without qualification" [29, I.15, 34b7–8].[72] While the role this clarification is meant to play in Aristotle's own treatment of syllogisms is not entirely clear,[73] Kilwardby employs this distinction primarily in order to dismiss apparent counterexamples to syllogisms that contain one assertoric and one modal premise. For example Kilwardby observes that Barbara LXL, a mood Aristotle endorses, suffers from this seeming counterexample:

> Of necessity every man is an animal;
> everything white is a man (let it be so);
> but not of necessity everything white is an animal [35, pt. 1, p. 322.235–237 (*dub.* 7)]

His response is to deny that the premise "everything white is a man" is an assertoric statement of the right kind: This is merely true "as of now", whereas premises in the modal syllogistic ought to be true "*simpliciter*".

Kilwardby describes this restriction in terms of the "power" of the major premise to "appropriate" the minor so as to be a *simpliciter* assertoric [35, pt. 1, p. 322.232–274].[74] He argues that in Celarent with a necessity major and a contingency minor, the necessity major similarly "appropriates the minor to itself so that it is a natural contingency".[75] This restriction is necessary, since Celarent LQX suffers from the following counter-example:

> of necessity no stone is a man;
> it's contingent for everything moving to be a stone,
> but it's not the case that nothing moving is a man [35, pt. 1, p. 512.331–333 (*dub.* 5)]

This is an instance of Celarent LQX, a mood Aristotle apparently endorses at *Prior Analytics* I.16, 36a7. Yet while the premises of this argument are true, the conclusion need not be. Kilwardby's response is to add the stipulation that the minor premise be a natural contingency [35, pt. 1,

---

[72]μὴ κατὰ χρόνον ὁρίσαντας, οἷον νῦν ἢ ἐν τῷδε τῷ χρόνῳ, ἀλλ' ἁπλῶς

[73]For a different view, which does not take *simpliciter* assertorics to hold at all times, see Malink [18, pp. 234–237].

[74]For a discussion of Kilwardby's doctrine of appropriation as it relates to mixed necessity syllogisms, see Mendelsohn [20].

[75]*appropriat sibi minorem ut sit de contingenti nato* [35, pt. 1, p.526.582].

p.516.419–422 (*dub.* 5)]. An unrestricted assertoric conclusion *does* follow from a universal negative necessity and a universal affirmative natural contingency, as Kilwardby illustrates using the following example:

> Of necessity no musician is a log;
> it's contingent for every man to be a musician;
> so no man is a log [35, pt. 1, p. 528.589–591 (*dub.* 10)]

Here "every man is a musician" is taken to be a natural contingency, presumably with the idea that humans naturally develop a propensity for music so long as nothing impedes this development [35, pt. 1, pp. 154.385–156.386].[76] He also gives the following argument for the validity of this mood; however, the argument does little more than to re-state his definition of a natural contingency as a contingency with a dedicated cause:

> The reason is also clear as follows, that if it's impossible that some B is A when B contingently inheres in C according to a cause innate to it so that the minor is a natural contingency, it's clear that C will not be able to be A. And so there will be an unqualified assertoric conclusion of necessity, as is clear from the terms mentioned ('log, musician, man') [35, pt. 1, p. 530.620–625 (*dub.* 12)].

Kilwardby thus concludes that the syllogism Aristotle perfects, and which should be understood to exist on the basis of *Prior Analytics* 36a7–10, is Celarent $LQ_{nat}X_{simp}$. He alludes to, but does not spell out, an analogous argument intended to show, *mutatis mutandis*, that in Ferio with a necessity major and contingency minor, the minor premise must be taken to be a natural contingency (since this syllogism is perfected by Aristotle, on his interpretation, from Celarent $LX_{simp}L$).[77] And he gives an argument along similar lines for the conclusion that the Cesare

---

[76] On becoming a musician as a natural contingency see note 69.

[77] See [35, pt. 1, p. 500.135–142]; and cf. [35, pt. 1, pp. 516.423–426, 528.592–3] (note the plural *modis*). Kilwardby also claims in passing that the premises $ALeB$, $BQoC$ will yield the same conclusion as Ferio with a necessity major and a contingency minor since the minor premise can be converted "by opposite qualities" (i.e., complementary conversion [35, pt. 1, p. 500.142–146]). Here he seems either to forget his restriction of the Ferio syllogism to a natural contingency, or else to

syllogism with a necessity major and a contingency minor that Aristotle intends in *Prior Analytics* I.19 (38a16–21) is Cesare LQ$_{nat}$X$_{simp}$ (rather than Cesare LQX, LQ$_{ind}$X, etc.) [35, pt. 1, pp. 586.221–588.236]. It will be my object below, in formalizing Kilwardby's logic of natural contingencies, to show why these results do indeed follow given the way Kilwardby understands the meaning of natural contingency.

3 Formalizing Kilwardby's logic of natural contingency

The task of formalizing Kilwardby's logic in its entirety has been undertaken by Paul Thom [34]. This study goes beyond his previous work on Kilwardby's logic[78] in providing a model-theoretic semantics to capture the meaning Kilwardby takes syllogistic sentences to have, and meticulous accompanying proofs to show that this semantics captures Kilwardby's logical arguments and results. It is shown in detail how the background assumptions at work in Kilwardby's logic are stated and expounded in other works, such as Kilwardby's commentaries on Porphyry's *Eisagōgē*, *De Interpretatione* and the *Categories*. There are few parallels one can point to in contemporary scholarship on the history of logic that so adeptly combine first-rate textual scholarship with technical sophistication.

Following Kilwardby's claim in his commentary on the *De Interpretatione* and the *Prior Analytics* that terms correspond to concepts (*intellectus* or *rationes*), Thom interprets the sentences of Kilwardby's syllogistic logic in models that comprise a field of concepts ordered by various primitive intensional relations. In addition to a domain of individuals **D** and an associated interpretation function **I**$_D$, Thom's models thus also include a function **I**$_F$ which assigns terms of the language to elements of its conceptual field **F**. Some of these concepts are distinguished as *essences* (genera and species belonging to the category of substance), denoted by a set **E**. The field of concepts is structured by primitive relations

---

forget that he denies complementary conversion holds of natural contingencies. I will disregard this validity claim in what follows.

[78] See especially Thom [32] and Thom [31].

of inseparability ($\Leftarrow$), essential inseparability ($\Leftarrow_E$, which requires the subject to be an essence term), repugnance ($\Downarrow$) and denomination (AB). Collecting these elements into a tuple, we may identify a model with a structure

$$\mathcal{M} = \langle \mathbf{D}, \mathbf{I_D}, \mathbf{F}, \mathbf{E}, \mathbf{I_F}, \mathbf{R} \rangle$$

where $\mathbf{R}$ is a structure interpreting the primitive conceptual relations ($\Leftarrow$, $\Leftarrow_E$, $\Downarrow$, AB).

Necessities and contingencies are interpreted in terms of these primitive relations of repugnance and (essential) inseparability. It will not be necessary to go into the details of Thom's formalization of necessity and generic contingency and other modalities here.[79] The only important point, for our purposes, is that their truth conditions are given entirely in terms of $\mathbf{R}$ and $\mathbf{I_F}$. The domain of individuals and its associated interpretation function plays no direct role in their interpretation on Thom's reconstruction, making this an "intensional" interpretation of Kilwardby's modal logic.

Thom convincingly shows that this aligns with the way Kilwardby understands necessities and generic contingencies in syllogisms. However, this intensional framework is less well suited to capture Kilwardby's notion of a *simpliciter* assertoric. As we have seen, Kilwardby contrasts "as of now (*ut nunc*)" assertorics with *simpliciter* assertorics which are taken to hold at all times. As Thom interprets the distinction, the truth conditions of *ut nunc* assertorics are extensional (given in terms of $\mathbf{D}$ and $\mathbf{I_D}$), whereas the truth conditions of *simpliciter* assertorics are intensional (given in terms of $\mathbf{F}$ and $\mathbf{I_F}$). In particular, a universal affirmative *ut nunc* assertoric is true just if the subject term's extension is non-empty and included in the extension of the predicate term:

AaB is true in $\mathcal{M}$ iff $\mathbf{I_D}(B) \neq \emptyset \wedge \mathbf{I_D}(B) \subseteq \mathbf{I_D}A$

The corresponding *simpliciter* proposition, on the other hand, says on Thom's semantics that the concept associated with the subject term bears the intensional relation of inseparability to the predicate:

---

[79] For these, see Thom [33, chs. 5 and 6].

$AXaB$ is true in $\mathcal{M}$ iff $\mathbf{I_F}(A) \Leftarrow \mathbf{I_F}(B)$.[80]

Similarly, a universal negative assertoric says that the extensions of the two terms have an empty intersection, while the *simpliciter* assertoric says, in Thom's formalization, that the concept associated with the subject term bears the intensional incompatibility relation to the concept associated with the predicate term:

AeB is true in $\mathcal{M}$ iff $\mathbf{I_D}(B) \cap \mathbf{I_D}(A) = \varnothing$

$AXeB$ is true in $\mathcal{M}$ iff $\mathbf{I_F}(A) \Downarrow \mathbf{I_F}(B)$

The truth conditions of particular assertorics are the negations of the propositions of opposite quality, guaranteeing Aristotle's square of opposition for both *ut nunc* and *simpliciter* assertorics. In essence, the difference between the *ut nunc* and *simpliciter* truth conditions as formalized by Thom are thus (i) that the truth condition for an *ut nunc* assertoric is formulated in terms of extensional relations of set inclusion, exclusion and non-emptiness whereas the truth condition for a *simpliciter* assertoric is formulated in terms of Thom's special intensional relations, and (ii) the truth condition for the *ut nunc* assertoric is formulated in terms of the extensional interpretation function $\mathbf{I_D}$ (which, recall, maps to sets of individuals from the domain), whereas the *simpliciter* assertoric is formulated in terms of the intensional interpretation function $\mathbf{I_F}$ (whose values are concepts from the field of concepts $\mathbf{F}$).

What is noteworthy about this analysis for our purposes is that temporal notions play no role in it. This is perhaps surprising, since Kilwardby's distinction between *simpliciter* and *ut nunc* assertorics is quite explicitly a temporal one. In his commentary on *De Interpretatione*, Kilwardby glosses the distinction as follows:

> Something is taken to be *simpliciter* when something is predicated perpetually and incorruptibly, as in "man is an animal"; something is taken to be according to a time when something

---

[80] Like Thom [33], I use X to indicate a modality of *simpliciter assertoric* in a proposition (like AXaB), but to indicate regular assertoric in the name of a syllogism (like Barbara LXL). I use X_{simp} in the name of the syllogism to indicate that a premise is a simpliciter assertoric.

is predicated in time and variably, as "man is white".[81]

Kilwardby is contrasting predications which hold at some given time with those that hold at all times.[82] Now, it is plausibly a *sufficient* condition for an extensional predication to hold at any given time that a corresponding predication among concepts hold; at any rate, this is something Kilwardby seems to assume.[83] Further, Kilwardby also seems to hold that a *simpliciter* predication *entails* a relation of inseparability or repugnance between the concepts of the terms involved.[84] Hence, an intensional relation between terms does seem to be a necessary and sufficient condition for Kilwardby's notion of a *simpliciter* assertoric. However, that still does not mean that Kilwardby's distinction between *ut nunc* and *simpliciter* assertorics should be identified with the distinction between intensional and extensional predication, as it is in Thom's formalization. A more faithful representation of Kilwardby's position would introduce times into the models used to interpret syllogistic statements. We could then introduce bridging principles to guarantee that omnitemporal relations among extensions correspond with intensional

---

[81] Robert Kilwardby, *Notulae super librum Perihermeneias*, pt. 1, L.2 (*ad* 16a18): *esse simpliciter opponitur quando praedicatur aliquid perpetuum et incorruptibile, ut 'Homo est animal'; secundum tempus quando predicatur aliquid temporale et variabile, ut 'Homo est albus'*, quoted in Thom [33, p. 82n17]. My translation.

[82] I grant that the term "incorruptibly" (*incorruptibile*) suggests an intensional dimension to his understanding of the distinction as well. This is not surprising, since Kilwardby takes *simpliciter* assertorics to imply certain intensional conditions (see below). Nevertheless, it seems clear that the basic distinction here is between predications true at a particular time and those true throughout all time (the latter also being "incorruptible").

[83] For this assumption in Kilwardby, see Thom [33, pp. 88–90, 99] (especially postulates 4.1 and 4.2).

[84] See for instance [35, pt. 1, pp. 389–391], where Kilwardby claims that a *simpliciter* assertoric requires "that the concepts of the terms cohere or conflict (*quod rationes terminorum cohereant uel discohereant*)". Unlike Thom [33, pp. 88–89], I don't think this is meant by Kilwardby as a general definition of *simpliciter* predication. Kilwardby's point here is that *simpliciter* predication is not necessity in the *per se* sense required for necessary premises and conclusions in the modal syllogistic, *even though* they do entail a condition that might easily be confused with *per se* necessity. The assumption that a *simpliciter* predication implies an intensional relation among concepts is also evident in Kilwardby's association of *simpliciter* assertorics with habitudinal predication, which he does seem to take to be an intensional relation: See [35, pt. 1, p. 458.445–450] with Thom [33, p. 83].

relations among concepts.

I sketch such an alternative formalization below. This would be a pedantic complaint if *simpliciter* assertorics were only ever used in Kilwardby's logic in order to infer an intensional state of affairs from which further consequences may be derived. In that case, while the formalization proposed would still be closer to Kilwardby's own presentation, it would shed little light on the workings of Kilwardby's actual logic. The success of Thom's system in capturing Kilwardby's necessity and (generic) contingency syllogistic shows that this is indeed the *main* way that Kilwardby uses the notion of a *simpliciter* assertoric. However, this failure to take into account temporal notions in Kilwardby's logic becomes important when we try to interpret what Kilwardby has to say regarding *natural contingency*, as we will see.

3.1 Thom's formalization of Kilwardby on natural contingency

Let us begin with Thom's analysis of *potentiality*, which plays a key role in his analysis of natural contingencies (but not contingencies of other types). Thom includes in his language an operator on terms that produces, for every term A, a term A_p to be read as "the potential for A". Potentialities are governed by the sole axiom

If $\lambda \Leftarrow \mathbf{I_F}(A)$ and $\mathbf{I_F}(A_p) \Leftarrow \kappa$ and $\kappa \in \mathbf{E}$ then $\neg \lambda \Downarrow \kappa$ [34, p. 74]

That is, if something is inseparable from the actualization of a potentiality, and this potentiality is in turn inseparable from some essence, then that first thing must not be repugnant to that essence. This condition seems intended to capture Kilwardby's idea that potentialities, at least of the sort that syllogistic propositions concern, must be realizable: There can be no concepts which exist *merely potentially* but could never be realized.[85] Thom captures this idea in the above condition by stipulating that if a potentiality is inseparable from an essence concept (if, say, the capacity to become educated is inseparable from the concept human)

---

[85] Or if there are, they are not the concern of syllogistic logic: Cf. Thom [33, pp. 67–68].

then there cannot be a concept inseparable from the actualization of that potentiality (for instance, in this case, being literate) which is incompatible with the underlying essence kind. In other words, potentialities are never inseparable from concepts that would inherently preclude their actualization in an essence from which they are inseparable.

This notion of a potentiality forms the basis for Thom's interpretation of Kilwardby's natural contingencies. Thom only attempts to formalize universal affirmative natural contingencies, even though Kilwardby seems to allow particular affirmative necessities as well.[86] He proposes the following truth condition:

> $AQ_{nat}aB$ is true iff $AQaB$ is true and being A_p is inseparable from B, where B is an essence and A is a denominative [33, p. 152]

A universal affirmative natural contingency statement thus expresses the same as a regular (unampliated) universal affirmative contingency with the additional requirements that (i) the subject is an essence term and the predicate is a (substantivized) accident term (or "denominative"), and (ii) the *potentiality* to have the property expressed by the predicate is inseparable from the subject. Putting aside these restrictions on term kinds, this is to say that A holds naturally contingently of all B ($AQ_{nat}aB$) if, and only if (i) the contingency $AQaB$ holds, and (ii) being potentially A is inseparable from B.

This no doubt captures some aspects of what Kilwardby takes to be entailed by a natural contingency. A natural contingency is non-accidental: This is captured by the second clause, which requires that the potentiality be inseparable from the subject term. Kilwardby also takes natural contingencies to imply contingencies in the generic sense, which is ensured by the first clause. And Kilwardby does seem to take all naturally contingent properties to be at least possibly realizable, which is captured by Thom's axiom regarding potentialities.

---

[86] Kilwardby explicitly denies that there are negative natural contingencies [35, pt. 1, p.398.578–595], but he presupposes that there are particular affirmative natural contingencies when he endorses the validity of Ferio $LQ_{nat}X_{simp}$ [35, pt. 1, pp. 526.561–528.593 (*dub.* 10)].

This is, then, plausibly a necessary condition for natural contingency as Kilwardby understands it. Yet it is clearly not a sufficient condition: Beyond saying that a regular contingency is true, it says only that the potentiality associated with the predicate term is inseparable from the subject and must be realizable within it, and restricts the types of the relevant terms. Yet, as we have seen, Kilwardby holds something much stronger than this concerning natural contingencies. He claims not just that a naturally contingent property *can* always be realized in its subject; he holds that a natural contingency *always will* be realized in its subject so long as nothing obstructs its realization. For Kilwardby, the distinguishing feature of a natural contingency is that it describes a causal process which *occurs of necessity*[87] and as such *will necessarily terminate in actuality, given sufficient time to do so before the subject in which it occurs expires*.

This central element of Kilwardby's account is not captured by Thom's definition. Further, it is difficult to see how this could be captured in Thom's framework, since he takes the truth conditions of modal syllogistic propositions to be grounded in timeless relationships between concepts, whereas Kilwardby's analysis makes reference to individuals, time, and processes which occur in them over time.

In addition to these problems capturing Kilwardby's metaphysical analysis of natural contingency, it is also questionable whether Thom's interpretation adequately captures the syllogistic results involving natural contingency that Kilwardby sets out. As we have seen, Kilwardby holds that in a Celarent syllogism with a necessity major and a contingency minor, the minor must be a natural contingency in order to yield a *simpliciter* assertoric conclusion. As Thom interprets Kilwardby here, the conclusion of the Celarent syllogism must be interpreted not as an unrestricted assertoric with its usual semantics but rather as one whose meaning is the same as a (broad) possibility proposition, since the mood is reduced from Ferio LXL where the minor is a *simpliciter* assertoric. Thom introduces the notation X_M to mark this, and writes the relevant

---

[87]See [35, pt. 1, pp. 535–6], where "goes grey" is used as the example of the natural predicate: "If [the act of going grey] bespeaks the process of going grey then when there are men they are always and of necessity going grey [*si dicat motum in canitiem sic semper ex necessitate canescit homo cum est*]".

syllogism as Celarent $LQ_{nat}X_M$. But Kilwardby does not, as far as I can see, make the claim that the conclusion here is to be read merely as a possibility. He claims only that the conclusion is to be read as an unrestricted assertoric [35, pt. 1, pp. 516.423–424, 527.578–528.588]. And, besides, the inference with the stronger conclusion is intuitively valid, as Kilwardby points out [35, pt. 1, p.528.589–591 (*dub*. 10), discussed above]. If it is of necessity the case that nothing which develops grey hair is a log, and if it is naturally contingent for all men to develop grey hair, then it is always actually the case that no men are logs (and not merely possibly).

There are, then, three problems with Thom's formalization of natural contingency in Kilwardby. First, it fails to capture the metaphysics of natural contingency as Kilwardby understands it, in part because it fails to formalize Kilwardby's use of temporal and extensional notions in his account of what a natural contingency means. Thom claims that this metaphysical level of analysis is irrelevant to logic for Kilwardby,[88] but it is not clear why this should be so.[89] Second, Thom's formalization does not capture the central syllogistic result that Kilwardby endorses in connection with natural contingency (namely Celarent $LQ_{nat}X_{simp}$). Finally, there is the more minor problem that Thom's analysis fails to account for particular affirmative natural contingencies, which Kilwardby does seem to at least implicitly recognize. In the next section, I will consider in what way Thom's semantics must be enriched in order to more satisfactorily incorporate Kilwardby's analysis of natural contingency.

3.2 An alternative formalization of Kilwardby on natural contingency

How can we formalize Kilwardby's notion of natural contingency in a way that hews more closely to the meaning of natural contingency as Kilwardby understands it? As we have seen, Kilwardby's notion of natural contingency makes reference to individuals and to time. What makes

---

[88]"the science of logic does not need to refer to this metaphysical grounding" [33, p. 151].

[89]At any rate, Thom does not seem to have any qualms about referring to the metaphysical grounding of logic in order to develop the semantics for Kilwardby's logic of necessities, as we have seen.

the contingency that men go grey distinctively *natural*, for Kilwardby, is the fact that any particular man will eventually get grey hair if that man lives long enough.

It is not obvious how to formalize this condition in contemporary modal logic.[90] We will proceed by taking Kilwardby's statements about natural contingency as the basis for a model-theoretic semantics, using Thom's models as a basis. We expand these models to include a timeline **T**. For simplicity, we assume that time is non-branching and thus that **T** is a segment of the reals representing a timeline. Further, we will need to relativize the interpretation function over individuals, $\mathbf{I_D}$, to times:

- $\mathbf{I_D} : \mathbf{L} \times \mathbf{T} \to \mathcal{P}(\mathbf{D})$

where **L** is the set of terms in the language. We will also need a second extension function, $\mathbf{O_D}$, to represent the set of things that are F including those things which may not be actually existent:

- $\mathbf{O_D} : \mathbf{L} \times \mathbf{T} \to \mathcal{P}(\mathbf{D})$

We require that $\mathbf{I_D}(T,t) \subseteq \mathbf{O_D}(T,t)$ for all terms T and times t. $\mathbf{I_D}$ may now be thought of as a description of an "inner" domain that varies with time, making the logic generated by this model theory a free logic. Since we do not require that only actually existent individuals at t occur within the extension of the interpretation function $\mathbf{O_D}$, the logic generated by these models will be a free logic with positive semantics.[91]

We may now re-define *ut nunc* and *simpliciter* assertorics as follows, in line with the conclusions of the discussion above:

- AaB is true in $\mathcal{M}$ at t iff $\mathbf{I_D}(B,t) \neq \emptyset \wedge \mathbf{I_D}(B,t) \subseteq \mathbf{I_D}(A,t)$
- AeB is true in $\mathcal{M}$ at t iff $\mathbf{I_D}(B,t) \cap \mathbf{I_D}(A,t) = \emptyset$
- AiB is true in $\mathcal{M}$ at t iff AeB is not true in $\mathcal{M}$ at t
- AoB is true in $\mathcal{M}$ at t iff AaB is not true in $\mathcal{M}$ at t

- $AXaB$ is true in $\mathcal{M}$ at t iff $\mathbf{O_D}(B,t) \subseteq \mathbf{O_D}(A,t)$ is true in $\mathcal{M}$ for all t

---

[90] I am grateful to Stephen Read for drawing my attention to some of the difficulties involved. I provide a tentative suggestion in my concluding remarks below.

[91] On the classification of free logics see Nolt [23, sect. 3.2].

- $AXeB$ is true in $\mathcal{M}$ at t iff $\mathbf{O_D}(B,t) \cap \mathbf{O_D}(A,t) = \emptyset$ is true in $\mathcal{M}$ for all t
- $AXiB$ is true in $\mathcal{M}$ at t iff $\mathbf{O_D}(B,t) \cap \mathbf{O_D}(A,t) \neq \emptyset$ is true in $\mathcal{M}$ for all t
- $AXoB$ is true in $\mathcal{M}$ at t iff $\mathbf{O_D}(B,t) \nsubseteq \mathbf{O_D}(A,t)$ is true in $\mathcal{M}$ for all t

Here the *ut nunc* assertorics are unchanged from Thom's except for being relativized to a given time, and restricted to the inner domain [33, p. 99]. *Simpliciter* assertorics differ from plain assertorics according to these definitions in three ways: First, they require their condition to hold not just at one time but "perpetually"—that is, at all times.[92] Second, they concern also non-actual individuals at any given time. And third, the universal affirmatives do not have existential import.[93] In order to ensure the necessity and sufficiency of *simpliciter* predication for intensional relations, we may ascribe to Kilwardby the following *bridging principles* which are required to hold in any model:

1. $(\forall t \in \mathbf{T})(\mathbf{O_D}(A,t) \subseteq \mathbf{O_D}(B,t))$ if, and only if, $\mathbf{I_F}(A) \Leftarrow \mathbf{I_F}(B)$
2. $(\forall t \in \mathbf{T})(\mathbf{O_D}(A,t) \cap \mathbf{O_D}(B,t) = \emptyset)$ if, and only if, $\mathbf{I_F}(A) \Downarrow \mathbf{I_F}(B)$
3. $(\forall t \in \mathbf{T})(\mathbf{O_D}(A,t) \nsubseteq \mathbf{O_D}(B,t))$ if, and only if, $\mathbf{I_F}(A) \nLeftarrow \mathbf{I_F}(B)$
4. $(\forall t \in \mathbf{T})(\mathbf{O_D}(A,t) \cap \mathbf{O_D}(B,t) \neq \emptyset)$ if, and only if $\mathbf{I_F}(A) \nDownarrow \mathbf{I_F}(B)$

These principles may be compared with the "principle of plenitude":[94] They say, in effect, that there must be an intensional relation between terms underlying any eternal relation among extensions. However, they stipulate this equivalence with respect to extensions within the outer domain, not the inner domain.

Thom's remaining truth conditions can then be interpreted as they stand, with changes made only to relativize each truth condition to a given moment in time as with the *ut nunc* assertorics above. If we modify

---

[92] *perpetuum*, Robert Kilwardby, *Notulae super librum Perihermeneias*, pt. 1, L.2 (*ad* 16a18), quoted in Thom [33, 82n17].

[93] Cf. Thom [33, p. 100]. Thanks to Christophe Geudens for bringing my attention to issues of existential import here.

[94] See Hintikka [10]; for the prevalence of these assumptions in early scholastic philosophy, see Knuuttila [14] and Knuuttila [15], especially ch. 3.

the definition of semantic entailment in the obvious way,[95] then all of Thom's positive results concerning *simpliciter* assertorics will remain intact, since, given the above bridging principles, these entail (and are entailed by) the intensional conditions taken to be their truth conditions by Thom.[96]

This distinction between an inner and an outer domain allows us to capture what makes a contingency natural for Kilwardby. What makes "men go grey" a distinctively natural kind of contingency, on Kilwardby's analysis, is the satisfaction of the additional condition that any given man *would eventually go grey, were he to live a sufficiently long time*. In these models, we can capture this by the condition that every man is, at some point in the future when that man *may or may not exist*, part of the (outer) extension of "grey".[97] If you like: All men eventually do go grey, but this may occur after that man has ceased to exist.

That suggests the following truth-conditions for natural contingencies in these models:

- $\mathcal{M} \vDash AQ_{nat}aB$ iff (i) $\mathcal{M} \vDash AQaB$, and (ii) $(\forall t \in \mathbf{T})(\forall x \in \mathbf{D})(x \in \mathbf{O_D}(B,t) \supset (\exists t' \geq t)(x \in \mathbf{O_D}(A,t')))$
- $\mathcal{M} \vDash AQ_{nat}iB$ iff (i) $\mathcal{M} \vDash AQiB$ and (ii) $(\forall t \in \mathbf{T})(\exists x \in \mathbf{D})(x \in \mathbf{O_D}(B,t) \wedge (\exists t' \geq t)(x \in \mathbf{O_D}(A,t')))$

That is, a natural contingency requires a corresponding contingency to be true (where this is understood, as per Thom's analysis, as an intensional condition), and *also* for there to be a future time when the potentiality is realized (at which time the subject in question may or may not exist!). The condition is required to hold at all times so as to capture the fact that we are talking about *natural* processes: If it is *natural* for some or all men to go grey, we assume that this is a process that applies to men at all times (even if, as Kilwardby stresses, the process does not always reach completion). We define only positive statements, in line with Kilwardby's view that "natural contingencies [...] are not much

---

[95]$\mathcal{M} \vDash_t \phi$ iff ϕ is true in $\mathcal{M}$ at t; $\mathcal{M} \vDash_t \{\phi_1,\ldots,\phi_n\}$ iff $\mathcal{M} \vDash_t \phi_1,\ldots,\mathcal{M} \vDash_t \phi_n$; and $\Gamma \vDash \phi$ iff for all models $\mathcal{M}$, for all times t, if $\mathcal{M} \vDash_t \Gamma$, then $\mathcal{M} \vDash_t \phi$.

[96]Some additional modifications may be needed to this definition in order to stop it from validating unwanted *invalidity* results. I will not consider this issue here.

[97]I am assuming here that a man exists if, and only if, that man is alive.

mentioned under a negation but always positively" [35, pt. 1, pp. 578–588 (*dub.* 9)].[98]

No modifications will be required to Thom's truth conditions for other syllogistic modalities which, as mentioned, are formulated in terms of primitive conceptual relations over $\mathbf{I_F}$.[99] The only assumption we need to make about the logic of necessities is that negative universal necessities imply *simpliciter* (temporally unrestricted) assertorics defined here:

If $\mathcal{M} \vDash_t ALeB$ then $\mathcal{M} \vDash_t AXeB$ (Necessity Implies Omnitemporality)

Kilwardby endorses this principle, although he denies the converse [35, pt. 1, p. 514.387–392].[100] Similar requirements for a, i and o propositions could be introduced but I will not do so here since they are also not needed for the results proven below.[101]

In order to simplify the representation of natural contingencies and to capture Kilwardby's other claims about this modality, we introduce two new term-level operators. Let $\dot{A}$ be read as "has reached the completion of a natural process which naturally terminates in A", and define $\tilde{A}$ so as to mean "is naturally undergoing a process which culminates in having A, or has reached the completion of this process." I will refer to a term of the form $\dot{A}$ as a result-term and one of the $\tilde{A}$ as a process-term. In order to capture the intended meaning of these operators in our models, we specify the following three axioms concerning result- and process-terms:

---

[98] I dispense here with Thom's requirements that the subject be an essence term and the predicate a denominative, since it is my goal to find a formalization that captures the logical results Kilwardby holds in connection with natural contingency, and these extra conditions are not required for any of these results as far as I can see. Nevertheless, it may be a part of Kilwardby's conception of natural contingency that the terms be restricted in this way, in which case these conditions could be added to the truth conditions above.

[99] I do not list these truth conditions, since they are not needed for the results derived here. They are given in Thom [33, pp. 126–127, 153–154].

[100] This also holds on Thom's definition of *simpliciter* assertorics, which I have stipulated mine to be logically equivalent to. Hence this will hold on Thom's semantics, with the above modifications.

[101] These would also need to be modified because Kilwardby takes the subject terms of some modal propositions to be unampliated: See Thom [33, p. 120].

Axiom 1 $O_D(\tilde{G}, t) \subseteq O_D(\dot{G}, t')$ for some $t' \geq t$ [**natural processes reach completion**]

Axiom 2 $O_D(\dot{G}, t) \subseteq O_D(\dot{G}, t')$ for every $t' \geq t$ [**completion of processes cannot be undone**]

Axiom 3 $O_D(\dot{G}, t) \subseteq O_D(\tilde{G}, t')$ for every $t' \leq t$ [**all natural results are the result of a natural process**]

Axiom 1 says that, if there is a time at which x is in the extension of $\tilde{A}$, there is a posterior time at which it is in the extension of $\dot{A}$. For instance, if there is a time at which a man is going grey, there is a posterior time at which that man is grey. Note that axiom 1 does *not* specify that x still be an existing individual at t'. This is important, since it means that axiom 1 does not say (in the case of men going grey) that every man who is in the process of going grey will live to actually be grey. The posterior time may be one at which the individual x is no longer in existence, in which case the axiom only requires that the individual *would be grey, were he still alive.*[102]

Axiom 2 specifies that the results of natural processes are permanent: If something gets grey hair as a result of this process, then it will always have naturally grey hair. Hair treatments may mask the natural colour of one's hair, but they do not change the natural colour of one's hair. In other words, the effect of having completed a natural process cannot be undone.

Finally, axiom 3 says that natural results do not come out of nowhere: Each individual with a natural result G (e.g. having grey hair) either already had G or was in the process of getting it at all prior times.[103]

It can then be shown that:

- $\mathcal{M} \vDash_t \dot{A}Q_{nat}aB$ iff (i) $M \vDash_t \dot{A}QaB$, and (ii) $M \vDash_t \tilde{A}XaB$

- $\mathcal{M} \vDash_t \dot{A}Q_{nat}iB$ iff (i) $M \vDash_t \dot{A}QiB$ and (ii) $M \vDash_t \tilde{A}XiB$

---

[102] Readers who are comfortable with possibilia might prefer to say that the now-merely-possible-man turns grey.

[103] Here, for simplicity's sake, I make the admittedly artificial assumption that natural processes extend back infinitely in time, so that, for example, each possible man was already going grey before that man was born.

In other words, the completion of a natural process is predicated naturally contingently, if, and only if, the corresponding (plain) contingency holds and it is always the case that subjects of that sort naturally have that property or are in the process of receiving it. This is proven in the appendix (4.1).

Now, given the bridging principles stated above, we may infer from a temporally unrestricted assertoric that a primitive conceptual relation holds. It follows that clause (ii) may be rewritten to give the following truth conditions for natural contingencies:

- $\mathcal{M} \vDash_t \dot{A}Q_{nat}aB$ iff (i) $M \vDash_t \dot{A}QaB$, and (ii) $\mathbf{I_F}(\tilde{A}) \Leftarrow \mathbf{I_F}(B)$
- $\mathcal{M} \vDash_t \dot{A}Q_{nat}iB$ iff (i) $M \vDash_t \dot{A}QiB$ and (ii) $\mathbf{I_F}(\tilde{A}) \not\Leftarrow \mathbf{I_F}(B)$

For a result-term, a natural contingency is therefore equivalent to the holding of a regular contingency [35, pt. 1, pp. 133–134] where there is also a necessary intensional relationship between the subject concept and the concept of the predicate's process or completion. This result captures Kilwardby's view of the relationship between natural contingency and necessity, according to which a natural contingency is closely associated with a necessary predication without being equivalent to one [35, pt. 1, pp. 535–536].[104]

It is also easy to see that, on these semantics, the version of complementary conversion Kilwardby states for natural contingencies does in fact hold. This is proven in appendix (4.2). Further, under the plausible assumption that true negative *per se* necessities are true at all times, it can be shown that these agree with Kilwardby's syllogistic results for natural contingencies. The central result, as we have seen, is Celarent LQ$_{nat}$X$_{simp}$, where the conclusion is to be understood as temporally unrestricted.

We can also show that Celarent LQ$_{nat}$X$_{simp}$ holds, so long as we consider predicates that describe the completion of natural processes, as Kilwardby

---

[104]The relevant necessity will not be the *per se* type of necessity Kilwardby takes to be at issue in the modal syllogistic, since it neither expresses a genus-species relation in the category of substance nor a relationship between two abstract terms (like "white is coloured"); it will, however, plausibly be what Kilwardby elsewhere calls a necessity *per accidens*: See Thom [33, pp. 113–119].

does.[105] Suppose that in some model $\mathcal{M}$, for some time t, (i) $\mathcal{M} \vDash_t \dot{O}Le\dot{G}$ and (ii) $\mathcal{M} \vDash_t \dot{G}Q_{nat}aM$. For vividness, read O as "has orange hair", G as "has grey hair" and M as "is a man"; then these say that necessarily nothing which has naturally obtained orange hair has naturally obtained grey hair and it is naturally contingent for all men to naturally obtain grey hair. Now, if it is necessary for nothing naturally grey to be naturally orange, and if all men naturally contingently have grey hair, then no man ever has naturally orange hair, since there would then need to be a future time at which the possible or actual man still has orange hair (since natural processes cannot be undone; axiom 2) but at which time he also has naturally grey hair (since natural processes reach completion; axiom 1). A more formal proof is given in the appendix (4.4), as well as an analogous proof for Ferio $LQ_{nat}X_{simp}$. Since the third syllogistic mood involving natural contingency that Kilwardby recognizes, Cesare $LQ_{nat}X_{simp}$, reduces to Celarent by conversion, this suffices to capture Kilwardby's syllogistic logic of natural contingency.

3.3 Concluding remarks: Natural contingency in Kilwardby and beyond

While less developed than his theory of *per se* necessity and possibility, Kilwardby's remarks about natural contingency provide enough detail to give the outlines of an interesting logical system. It is probably still correct to describe Kilwardby's logic as an intensional system, but the foregoing shows that there is at least one type of modality where extensional and temporal notions are important to the way Kilwardby understands the meaning of modal language.

I am not aware of explorations of the logic of natural processes and

---

[105] I will not attempt to show that the mood is unrestrictedly valid. Kilwardby's theory of natural contingency only motivates considering the result for predicate terms which represent the completion of natural processes, since this is the reading of the predicate required in order to make a natural contingency true. I therefore restrict the minor premise to a result-term. I make the further assumption that the other term in the necessity is also a result-term. Note, however, that the $\tilde{}$ and $\dot{}$ operators allow us to express this as a formally valid argument.

results in contemporary logic.[106] A project for future research would be to consider the logic of a language of temporal predicate logic that included process- and result-forming operators for predicates, and explore the modalities that naturally arise in it. If we abstract from Kilwardby's reliance on a syllogistic framework and consider only his treatment of natural contingency as a modality, then a natural first step in formalizing a modal operator of natural contingency along the lines of Kilwardby's analysis in a predicate-logical language might be to define, for atomic sentences:

$$\Diamond_{nat} G(a,t) \equiv_{def} \Diamond G(a,t) \land \Diamond \neg G(a,t) \land (\exists t')(\forall t'' \geq t')(E!(a,t'') \to G(a,t''))$$

That is, it is naturally contingent that Ga at some given time t if, and only if, it is at that time possible that Ga and possible that $\neg Ga$ and there is a future time after which a will be G, so long as a exists (here $E!$ is an existence predicate). I leave it to future research to consider how, or whether, this could be generalized beyond the case of atomic propositions.

4 Appendix: Supplementary technical results

4.1 Natural contingency truth conditions with process- and result-terms

We show that given the general definition of natural contingency

$$\mathcal{M} \vDash_t AQ_{nat}aB \text{ iff (i) } \mathcal{M} \vDash_t AQaB, \text{ and (ii) } (\forall x \in \mathbf{D})(\forall t \in \mathbf{T})(x \in \mathbf{O_D}(B,t) \supset (\exists t' \geq t)(x \in \mathbf{O_D}(A,t')))$$

it follows that

$$\mathcal{M} \vDash_t \dot{A}Q_{nat}aB \text{ iff (i) } \mathcal{M} \vDash_t \dot{A}QaB, \text{ and (ii) } \mathcal{M} \vDash_t \tilde{A}XaB$$

---

[106] The logic of *processes* has been explored [see e.g. 25, pp. 155–169], but I know of no explorations of the logic of processes that are distinctively *natural* and their natural outcomes. For a formal analysis of the metaphysics of some closely related notions, however, see Freddoso [9].

The first clause of the definiens obviously agrees with the first clause of the general definition. It needs to be shown that the second clause agrees with the general definition, i.e. that $(\forall x \in \mathbf{D})(\forall t \in \mathbf{T})(x \in \mathbf{O_D}(B,t) \supset (\exists t' \geq t)(x \in \mathbf{O_D}(\dot{A},t'))) \Leftrightarrow M \vDash_t \tilde{A}XaB$. We argue as follows:

Right to left: Suppose that $M \vDash_t \tilde{A}XaB$. Then $\mathbf{O_D}(B,t') \subseteq \mathbf{O_D}(\tilde{A},t')$ for all t' by the *simpliciter* assertoric truth condition. By axiom 1 there is a $t'' \geq t'$ such that $\mathbf{O_D}(B,t') \subseteq \mathbf{O_D}(\dot{A},t'')$, for any t'. That is to say, $(\forall x \in \mathbf{D})(\forall t \in \mathbf{T})(x \in \mathbf{O_D}(B,t) \supset (\exists t' \geq t)(x \in \mathbf{O_D}(\dot{A},t')))$

Left to right: Suppose that $(\forall x \in \mathbf{D})(\forall t \in \mathbf{T})(x \in \mathbf{O_D}(B,t) \supset (\exists t' \geq t)(x \in \mathbf{O_D}(\dot{A},t')))$. Take any given time r. If there is no $a \in \mathbf{O_D}(B,r)$, then the conditional is trivially satisfied. If there is such an a, then $a \in \mathbf{O_D}(\dot{A},s)$ for some time $s \geq r$. Now by axiom 3, $a \in \mathbf{O_D}(\tilde{A},r)$. Since a is an arbitrary member of $\mathbf{O_D}(B,r)$, it follows that $\mathbf{O_D}(B,r) \subseteq \mathbf{O_D}(\tilde{A},r)$. Since r is an arbitrary time it follows that, for all times t', $\mathbf{O_D}(B,t') \subseteq \mathbf{O_D}(\tilde{A},t')$. This is the truth condition for $\tilde{A}XaB$. Hence, $M \vDash_t \tilde{A}XaB$.

The proof for the particular affirmative is analogous.

4.2 Q_{nat}/M conversion

These follow trivially given the rules of (i) complementary conversion ($AQaB \dashv\vDash AQeB$ and $AQiB \dashv\vDash AQoB$) (ii) Q-M weakening ($AQ \circ B \vDash AM \circ B$ for $\circ = a, e, i, o$)

1. $AQ_{nat}aB \vDash AMeB$. Proof: Suppose $AQ_{nat}aB$ is true in a model $\mathcal{M}$. By the truth condition for $AQ_{nat}aB$, $AQaB$ holds in $\mathcal{M}$. By complementary conversion, $AQeB$ holds in $\mathcal{M}$. Then by Q-M weakening, $AMeB$ holds in $\mathcal{M}$.

2. $AQ_{nat}iB \vDash AMoB$. Proof: Suppose $AQ_{nat}iB$ is true in a model $\mathcal{M}$. By the truth condition for $AQ_{nat}iB$, $AQiB$ holds in $\mathcal{M}$. By complementary conversion, $AQoB$ holds in $\mathcal{M}$. Then by Q-M weakening, $AMoB$ holds in $\mathcal{M}$.

4.3 Ferio $LQ_{nat}X_{simp}$

We will show that $\dot{O}Le\dot{G}, \dot{G}Q_{nat}iM \vDash \dot{O}XoM$.

Suppose that in some model $\mathcal{M}$, for some time t, (i) $\mathcal{M} \vDash_t \dot{O}Le\dot{G}$ and (ii) $\mathcal{M} \vDash_t \dot{G}Q_{nat}iM$. Assume for reductio that the temporally unrestricted conclusion were false at t. Then there is a time s such that $\mathbf{O_D}(M,s) \subseteq \mathbf{O_D}(\dot{O},s)$. By (ii), $\mathbf{O_D}(M,s) \subseteq \mathbf{O_D}(\tilde{G},s)$. Let $a \in \mathbf{O_D}(M,s)$ (we are guaranteed such an individual by (ii)). Then $a \in \mathbf{O_D}(\dot{O},s)$ and $a \in \mathbf{O_D}(\tilde{G},s)$. By axioms 1 and 2, however, there is a time $v \geq s$ such that $a \in \mathbf{O_D}(\dot{G},v)$ and $a \in \mathbf{O_D}(\dot{O},v)$. Since Necessity Implies Omnitemporality, however, $\mathbf{O_D}(\dot{G},v) \cap \mathbf{O_D}(\dot{O},v) = \emptyset$, a contradiction. We therefore reject the assumption for reductio and conclude that a simpliciter assertoric conclusion holds. That is, Ferio LQ$_{nat}$X$_{simp}$ is valid for completion predicates.

4.4 Celarent LQ$_{nat}$X$_{simp}$

We will show that $\dot{O}Le\dot{G}, \dot{G}Q_{nat}aM \vDash \dot{O}XeM$.

Suppose that in some model $\mathcal{M}$, for some time t, (i) $\mathcal{M} \vDash_t \dot{O}Le\dot{G}$ and (ii) $\mathcal{M} \vDash_t \dot{G}Q_{nat}aM$. Assume for reductio that the temporally unrestricted conclusion were false. Then by the truth condition for a universal negative *simpliciter* assertoric, there is a time u and an individual a for which $a \in \mathbf{O_D}(\dot{O},u)$ and $a \in \mathbf{O_D}(M,u)$. By (ii), however, $a \in \mathbf{O_D}(\tilde{G},u)$, and so by axiom 1 there is a future time v such that $a \in \mathbf{O_D}(\dot{G},v)$. At this time, however, we also have $a \in \mathbf{O_D}(\dot{O},v)$ by axiom 2. By Necessity Implies Omnitemporality, however, (i) implies that $\mathbf{O_D}(\dot{G},v) \cap \mathbf{O_D}(\dot{O},v) = \emptyset$, a contradiction. We therefore reject the assumption for reductio and conclude that a simpliciter assertoric conclusion holds. That is, Celarent LQ$_{nat}$X$_{simp}$ is valid for completion predicates.

References

[1] F. Alessio, *Lambert of Auxerre. Logica (Summa Lamberti)*. La Nuova Italia, Florence, 1971.

[2] J. Barnes, 'Sheep Have Four Legs' in M. Bonelli (ed.) *Logical Matters: Essays in Ancient Philosophy II*. Clarendon Press, Oxford, 2012, pp. 346–353.

[3] A. Becker, *Die Aristotelische Theorie der Möglichkeitsschlüsse*. Wissenschaftliche Buchgesellschaft, Darmstadt, 2nd ed., 1933.

[4] A. Borgnet, *B. Alberti Magni Opera Omnia*, vol. 1. Vives, Paris, 1890.

[5] D. Cannone, 'Le «Notule Libri Posteriorum» di Robert Kilwardby: il commento ad Analitici Posteriori I, 4, 73a34-b24' in *Documenti e studi sulla tradizione filosofica medievale*, 2002, 13, pp. 1–70.

[6] A. de Libera, 'Les summulae dialectices de Roger Bacon: I-II. De termino, de enuntiatione' in *Archives d'histoire doctrinale et littéraire du Moyen Âge*, 1986, 53, pp. 139–289.

[7] L. M. de Rijk, ed., *Logica Modernorum: A Contribution to the History of Early Terminist Logic, vol. 2: The Origin and Early Development of the Theory of Supposition*. Koninklijke Van Gorcum & Comp., Assen, 1967.

[8] S. Ebbesen, 'Analyzing syllogisms or Anonymus Aurelianensis III– the (presumably) earliest extant Latin commentary on the *Prior Analytics*, and its Greek model' in *Cahiers de l'Institut du Moyen-Âge Grec et Latin*, 1981, 37, pp. 1–20.

[9] A. Freddoso, 'The Necessity of Nature' in *Midwest Studies in Philosophy*, 1986, 11, pp. 215–242.

[10] J. Hintikka, 'Necessity, Universality, and Time in Aristotle' in J. Barnes, M. Schofield, and R. Sorabji (ed.) *Articles on Aristotle, vol. 3: Metaphysics*. Duckworth, London, 1979.

[11] J. Hintikka, *Time and Necessity: Studies in Aristotle's Theory of Modality*. Clarendon Press, Oxford, 1973.

[12] K. Jacobi, *Die Modalbegriffe in den logischen Schriften des Wilhelm von Shyreswood*. Brill, Leiden, 1980.

[13] K. Jacobi, 'Kontingente Naturgeschenisse' in *Studia mediewistyczne*, 1977, 18, pp. 3–70.

[14] S. Knuuttila, 'Medieval modal theories and modal logic' in D. M. Gabbay and J. Woods (ed.) *Handbook of the History of Logic, vol. 2*. Elsevier, Amsterdam, 2008, pp. 505–578.

[15] S. Knuuttila, *Modalities in Medieval Philosophy*. Routledge, London, 1993.

[16] H. Lagerlund, *Modal Syllogistics in the Middle Ages*. Brill, Leiden, 2000.

[17] H. G. Liddell and R. Scott, *A Greek-English Lexicon*. Clarendon Press, Oxford, 1940. Revised and augmented throughout by Sir Henry Stuart Jones with the assistance of Roderick McKenzie.
[18] M. Malink, *Aristotle's Modal Syllogistic*. Harvard University Press, Cambridge, Mass., 2013.
[19] S. McCall, *Aristotle's Modal Syllogisms*. North-Holland Pub. Co., Amsterdam, 1963.
[20] J. Mendelsohn, 'Term kinds and the formality of Aristotelian modal logic' in *History and Philosophy of Logic*, 2017, 38, pp. 99–126.
[21] L. Minio-Paluello, *Boethius translator Aristotelis secundum 'recensionem Carnutensem'. Analytica priora (recensio Carnutensis – excerpta)*, vol. 3.1–4 of *Aristoteles Latinus*. Desclée De Brouwer, Bruges-Paris, 1962, pp. 143–191.
[22] L. Minio-Paluello, *Boethius translator Aristotelis. Analytica priora (recensio Florentina)*, vol. 3.1–4 of *Aristoteles Latinus*. Desclée De Brouwer, Bruges-Paris, 1962, pp. 5–139.
[23] J. Nolt, 'Free Logic' in E. N. Zalta (ed.) *The Stanford Encyclopedia of Philosophy*. The Metaphysics Research Lab, Stanford, Winter 2020.
[24] T. Parsons, *Articulating Medieval Logic*. Oxford University Press, Oxford, 2014.
[25] N. Rescher and A. Urquhart, *Temporal Logic*. Springer, Vienna, 1971.
[26] W. D. Ross, *Aristotle's Prior and Posterior Analytics*. Clarendon Press, Oxford, 1949.
[27] J. F. Silva, 'Robert Kilwardby' in H. Lagerlund (ed.) *Encyclopedia of Medieval Philosophy*. Springer, Dordrecht, 2011, pp. 1148–1153.
[28] R. Smith, *Aristotle. Prior Analytics*. Hackett, Indianapolis, 1989.
[29] G. Striker, *Aristotle. Prior Analytics. Book 1*. Clarendon Press, Oxford, 2009.
[30] G. Striker, 'Notwendigkeit mit Lücken. Aristoteles über die Kontingenz der Naturvorgänge' in *Neue Hefte für Philosophie*, 1985, 24–25, pp. 146–164.
[31] P. Thom, *Logic and Ontology in the Syllogistic of Robert Kilwardby*. Brill, Leiden, 2007.

[32] P. Thom, *Medieval Modal Systems: Problems and Concepts.* Ashgate, Hampshire, 2003.
[33] P. Thom, 'Robert Kilwardby on Syllogistic Form' in P. Thom and H. Lagerlund (ed.) *A Companion to the Philosophy of Robert Kilwardby.* Brill, Leiden, 2013, pp. 131–162.
[34] P. Thom, *Robert Kilwardby's Science of Logic. A Thirteenth-Century Intensional Logic.* Brill, Leiden, 2019.
[35] P. Thom and J. Scott, *Robert Kilwardby. Notuli libri Priorum. Critical edition with translation, introduction and indexes.* Oxford University Press, Oxford, 2016.
[36] C. T. Thörnqvist, *'Anonymus Aurelianensis III' in Aristotelis* Analytica priora*: Critical Edition, Introduction, Notes, and Indexes.* Brill, Leiden, 2014.
[37] C. T. Thörnqvist, 'The 'Anonymus Aurelianensis III' and the Reception of Aristotle's *Prior Analytics* in the Latin West' in *Cahiers de l'Institut du Moyen-Âge grec et latin*, 2010, 79, pp. 25–41.
[38] M. Wallies, *Alexandri in Aristotelis Analyticorum Priorum librum I commentarium,* vol. *2.1* of *Commentaria in Aristotelem Graeca.* G. Reimer, Berlin, 1883.

Part II

Modern age logic

Caramuel's Pentagon of Opposition and his Vindication of the Principle *Ex Contradictorio Quodlibet*

Wolfgang Lenzen
University of Osnabrück
lenzen@uos.de

1 Introduction

Juan Caramuel y Lobkowitz (1606-1682) was an enigmatic person with many interests and talents. Already as a child he solved difficult mathematical problems, and later on he worked as a philosopher, theologian, astronomer, physicist, and mathematician. He is said to have published more than 260 books, of which, however, "only little was of lasting importance"[1]. His main *logical* works comprise

- *Rationalis et realis philosophia* of 1642,
- *Theologia rationalis* of 1654,
- *Herculis logici Labori Tres* of 1655,
- *Moralis seu Politica logica* of 1680.

In the standard works on the history of logic, Caramuel is either completely ignored or grossly underestimated. Thus, he isn't mentioned at all in Kneale's *The Development of Logic*, while in Risse's survey of *The Logic of Modern Age* he is discredited as "one of the weirdest thinkers

---

[1]Quoted according to the German Wikipedia article "Juan Caramuel y Lobkowitz", online access on September 29, 2020: "Caramuel veröffentlichte – laut der Zählung von Jean Noel Paquot – nicht weniger als 262 Bücher über Grammatik, Dichtung, Rhetorik, Mathematik, Astronomie, Physik, Politik, Kirchenrecht, Logik, Metaphysik und Theologie. Allerdings behielt kaum etwas davon dauerhafte Bedeutung."

whose ideas were full of wit rather than correct"[2]. It was not before the last third of the 20th century that a certain rehabilitation of Caramuel's logic took place. In particular Czech authors like Karel Berka, Petr Dvořák, and Stanislav Sousedík investigated Caramuel's "relational logic" which deals with *doubly quantified propositions* of the structure

DQ Every x/no x/some x
    stands/doesn't stand in relation R(x,y)
    to every y/to no y/to some y.[3]

Caramuel's most complete discussion of the logical relations that exist between such propositions is based on the example '[Person] x errs in [issue] y'. Schema DQ gives rise to 3*2*3 = 18 different propositions, ten of which, however, turn out to be equivalent to one of the following eight irreducible propositions:

DQ 1	Everybody errs in everything	$\forall x \forall y R(x,y)$
DQ 2	Somebody errs in everything	$\exists x \forall y R(x,y)$
DQ 3	Everybody errs in something	$\forall x \exists y R(x,y)$
DQ 4	Somebody errs in something	$\exists x \exists y R(x,y)$
DQ 5	Somebody doesn't err in everything	$\exists x \exists y \neg R(x,y)$
DQ 6	Everybody in something doesn't err	$\forall x \exists y \neg R(x,y)$
DQ 7	Somebody doesn't err in anything	$\exists x \forall y \neg R(x,y)$
DQ 8	Everybody doesn't err in anything	$\forall x \forall y \neg R(x,y)$

Fig. 1: Doubly quantified propositions

As the formulas on the right-hand side of this table show, '$\forall x$' and '$\exists y$' are here used to symbolize the quantifiers 'for every x' and 'for at least one y', respectively. Furthermore, in what follows, the following symbols are used:

- $\neg \alpha$ for the *negation* of a proposition α
- $\alpha \wedge \beta$ for the *conjunction* of two propositions α, β

---

[2]Cf. [22, pp. 351-354] Risse disqualifies Caramuel's logic as "abounding with idiosyncrasies", and he disregards his main work simply because of the inappropriateness of the title *Theologia rationalis*.

[3]Cf. in particular [23], [1], and [11]; unfortunately, all these papers are in Czech; the only exception is Dvořák's [12].

- $\alpha \vee \beta$ for the *disjunction* of two propositions α, β
- $\alpha \Rightarrow \beta$ for the relation of logical consequence between α and β.

Caramuel himself arranged propositions DQ 1-8 in the following way which somehow reminds of a double Square of Opposition: [5, p. 413]

Omnis errat in nullo	
Aliquis errat in nullo	Omnis in aliquo non errat
Aliquis in aliquo non errat	

Nullus errat in nullo	
Aliquis non errat in nullo	Nullus in aliquo non errat
Aliquis non in aliquo non errat	

Fig. 2: *Caramuel's arrangement of the doubly quantified propositions*

As was explained in [15], the propositions can be re-arranged in accordance with Caramuel's own explanations so as to form an *octagon of opposition* (see Fig. 3 below). In this diagram light arrows symbolize *subalternations*, i.e., logical implications, while dark lines connect propositions which are negations, i.e., *contradictories*, of each other.

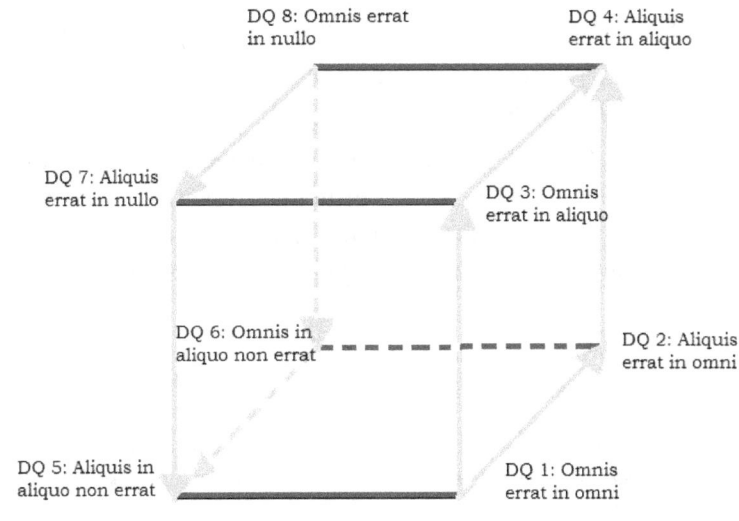

Fig. 3: *Caramuel's octagon (or "cube") of opposition*

193

The present paper, however, shall not be concerned with this *octagon* but rather with a *pentagon* of opposition which Caramuel obtained by adding to the traditional *square* of opposition of the four categorical forms

(A) Every S is P
(E) No S is P
(I) Some S is P
(O) Some S isn't P

the fifth form

(Y) Not every S is P.[4]

Before dealing with this issue in section 3, however, let us briefly touch upon another important discovery, namely:

2   Caramuel's Vindication of the Principle 'Ex contradictorio quodlibet'

Medieval logicians were well acquainted with the principles 'Ex impossibili quodlibet' and 'Necessarium ex quodlibet' according to which from an *impossible* proposition α, each other proposition β follows, while a *necessary* proposition β follows from each other proposition α:

EIQ   If α is impossible, then $(\alpha \Rightarrow \beta)$
NEQ   If β is necessary, then $(\alpha \Rightarrow \beta)$.[5]

E.g., in "Lesson 55" of his *Notule libri Priorum* written around 1240, Robert Kilwardby considers these principles together with two quite strange instantiations. On the one hand, NEQ is used (or better: abused) to infer that "if you are sitting then God exists, and if you are not

---

[4]The use of the vowel 'Y' (besides the traditional 'A', 'E', 'I', and 'O') goes back to Blanché's 1953 paper [2].

[5]In his introduction to part II of [13], Jacobi refers to these principles as "*ex impossibili quidlibet*" and "*necessarium a quolibet*", respectively. He further explains that, in the 12th century, both principles were accepted, e.g., by the school of the "Parvipotani" while the rivalling school of "Melidunenses" defended the contrary principle "*nil ex falso sequitur*".

sitting then God exists, because the necessary follows from anything"[6]. On the other hand, EIQ is (ab)used to infer "If you are an ass, you are not an ass because anything follows from the impossible and the necessary follows from anything".[7]

For readers who are not so familiar with the history of logic let it be pointed out that 'You are an ass' was never meant as a personal insult. Rather, it's a medieval standard example of an "impossible" assertion because the addressee of any assertion is a human being. But 'No human is an ass' (with 'ass' naming a biological species) is analytically true. Hence, for any person P, the proposition 'Person P is an ass' is (analytically) "impossible", while 'Person P is not an ass' is (analytically) "necessary". Therefore, at least in the opinion of many medieval logicians, the "necessary" proposition 'Person P is not an ass' follows from any other proposition, even from its own negation!

Furthermore, the vast majority of medieval philosophers not only believed in God but also thought that God's existence would be *provable*. Hence for them 'God exists' constituted a (theologically) "necessary" truth which, according to NEQ, follows from any other proposition, in particular both from the contingent assumption 'Person P is sitting' and from its negation, 'P is not sitting'. In a similar way, Stoic logicians believed that every piece of matter is ultimately composed of atoms. Thus, for them the proposition 'Atomic elements do exist' was (physically) "necessary" and hence it is entailed, in accordance with either NEQ or EIQ, by its own negation.[8]

---

[6]Cf. [14, p. 1140]: "[...] si tu sedes deus est et si tu non sedes deus est, quia necessarium ex quodlibet".

[7]Cf. [14, p. 1142]: "[...] 'Si tu es asinus, tu non es asinus' quia ex impossibili quidlibet et necessarium sequitur ad quodlibet." Kilwardby's examples challenge the validity of "Aristotle's Theses" which characterize so-called connexive logic. Aristotle's first thesis says that no proposition β is entailed both by a proposition α and by its negation, $\neg \alpha$; Aristotle's second thesis says that no proposition β is entailed by its own negation, $\neg \beta$. For a closer discussion of Kilwardby's position concerning "Aristotle's Theses" cf. [16].

[8]As Sextus Empiricus tells us in *Outlines of Pyrrhonism* ([3]), Diodorus, unlike Chrysippus, considered the paradoxical conditional 'If atomic elements do not exist, then atomic elements do exist' as sound. For a discussion of the Stoics' conceptions of conditionals cf. [17].

Now even if it is taken for granted that propositions like 'Person P is an ass', 'God doesn't exist', or 'Atomic elements do not exist' are in some sense *impossible*, one might still doubt whether they therefore *entail* every other proposition (and, in particular, their own negation). It might be objected that proposition α *entails*, i.e., *logically implies*, another proposition β only if there is a connection or relation between α and β such that, for *logical reasons*, β must be true if α is true. In other words, in order for $(\alpha \Rightarrow \beta)$ to hold, it must be *logically* impossible that α is true and yet β is false. *Other* kinds of impossibilities such as analytic impossibility, physical impossibility, or theological impossibility will not do!

On the background of such considerations, it seems reasonable to restrict the all too comprehensive principles EIQ and NEQ to the case where the antecedent is *logically impossible*, i.e., self-contradictory, or where the antecedent is *logically* necessary, i.e., tautological. The previous "Ex *impossibili* quodlibet" then turns into the narrower principle "Ex *contradictorio* quodlibet":

> ECQ From a contradictory proposition α (or from a contradictory set of propositions $\{\alpha_1, ..., \alpha_n\}$) any other proposition β follows.

In what follows, it suffices to concentrate on a simple instance of ECQ,

> ECQ1 $\{\alpha, \neg\alpha\} \Rightarrow \beta$ (or the variant $\alpha \wedge \neg\alpha \Rightarrow \beta$).

Already towards the end of the 12th century medieval logicians developed a *proof* of ECQ 1 which is based on elementary laws for the operators of disjunction and conjunction, in particular on the "disjunctive syllogism":

> DISJSYLL If a disjunction is true and if one of its parts is false, then the other part must be true.

According to [18], the proof was most likely invented by William of Soissons. At any rate, it is explicitly presented in Alexander Neckham's *De Naturis rerum* of around 1180:

> I wonder indeed that some men oppose the thesis that from a per se impossibility anything whatsoever follows. This may be confirmed with many reasons or brought to light with a few. For doesn't it follow that

if Socrates is a man, and Socrates is not a man, then Socrates is a man. But if Socrates is a man, then Socrates is a man or a stone; therefore, if Socrates is a man and Socrates is not a man, then Socrates is a man or a stone. But if Socrates is a man and Socrates is not a man, then Socrates is not a man; therefore, if Socrates is a man and Socrates is not a man, Socrates is a stone. By means of a similar deduction, it is proven that if Socrates is a man and Socrates is not a man, then Socrates is a crab, and so on for any other thing, for example a rose, a lily and anything else.[9]

Five hundred years later, in his *Leptotatos Latine Subtilissimus*, Caramuel discusses this proof with respect to the example "*Peter is running and Peter is not running. [...] Therefore, a circle has four corners*":

> In order to show that this conclusion follows infallibly, I make use of the common principles of logic accepted by everybody and argue thus. The proposition *Either Peter is running or a circle has four corners* is a disjunction, and therefore it is true. For it is clear that, for a disjunction to be true, it suffices that at least one of its parts is true. And in this disjunction the part *Peter is running* is true according to our supposition. [...] I continue further and argue thus: For each true disjunctive proposition there is a logically valid inference from the falsity of the one part to the truth of the other, because it cannot have two false parts (a proposition like *Peter is a stone or he is a tree*, both parts of which are false, is not a true disjunction). So, if one part is false, the other necessarily has to be true. If therefore in the true disjunctive proposition *Either Peter is running or a circle has four corners* the first part *Peter is running* is false (and this is so because, by our supposition,

---

[9]Cf. [20, pp. 288-289]: "Miror etiam quosdam damnare opinionem dicentium ex impossibili per se quodcunque sequi enuntiabile. Quod cum plurimis astrui queat rationibus, vel paucae prodeant in lucem. Nonne igitur si Sortes est homo, et Sortes non est homo, Sortes est homo? Sed si Sortes est homo, Sortes est homo vel lapis; ergo, si Sortes est homo, et Sortes non est homo, Sortes est homo vel lapis; sed si Sortes est homo, et Sortes non est homo, Sortes non est homo; ergo si Sortes est homo, et Sortes non est homo, Sortes est lapis. Consimili deductione, probabitur quod si Sortes est homo et Sortes non est homo, Sortes est capra, et ita de singulis rebus, puta rosa, lilio, et caeteris rebus." The above translation largely follows that of Martin in [19, p. 440].

it is true to say *Peter is not running*), the other part, namely *A circle has four corners*, necessarily has to be admitted.[10]

In modern times, adherents of a so-called "Paraconsistent logic" raised objections to ECQ. Thus, in a contribution to the *Stanford Encyclopedia of Philosophy*, [21], Graham Priest et al. explain:

> Contemporary logical orthodoxy has it that, from contradictory premises, anything follows. A logical consequence relation is *explosive* if according to it any arbitrary conclusion B is entailed by any arbitrary contradiction A, $\neg A$ (*ex contradictione quodlibet* (ECQ)). Classical logic, and most standard 'non-classical' logics too [...], are explosive. Inconsistency, according to received wisdom, cannot be coherently reasoned about. Paraconsistent logic challenges this orthodoxy. A logical consequence relation is said to be *paraconsistent* if it is not explosive. Thus, if a consequence relation is paraconsistent, then even in circumstances where the available information is inconsistent, the consequence relation does not explode into *triviality*.

The use of the word 'explosive' indicates that paraconsistent logicians are somehow *afraid* of a situation where the whole *world* (or at least the world of propositions) *collapses* because *everything* becomes *provable*. But this fear is unjustified. From the earliest beginnings of logic, *reductio ad absurdum* has always been acknowledged as an important tool for proofs. It says:

---

[10]Cf. [9, p. 215]: "*Petrus currit, & Petrus non currit* [...] Ergo [...] *Circulus habet quatuor angulos.* [...] Ut hanc Consequentiam ostendam infallibilem esse, utar communi et ab omnibus recepta Logica, & sic discurram. Haec, inquam, Propositio, *Vel Petrus currit, vel circulus habet quatuor angulos*, est Disjunctiva. Ergo & vera. Patet, quoniam ut disjunctiva vera sit, sufficit alteram in illa parte[m] esse veram. Et in hac illa pars *Petrus currit* ex suppositione vera est. [...] Pergo ulterius, & sic inquam. In omni Propositione Disjunctiva vera, a negatione alterius partis ad affirmationem alterius valet Consequentia; quia, cum non possit habere utramque partem falsam (Haec enim, *Vel Petrus est lapis, vel arbor*, cujus utraque pars est falsa, non erit Disjunctiva vera.) posito, quod altera pars falsa sit, altera necessario erit vera. Ergo in hac Propositione Disjunctiva vera, *Vel Petrus currit, vel circulus habet quatuor Angulos*, siquidem prima pars, *Petrus currit* est falsa; (est enim verum, ut praecipit Hypothesis, dicere *Petrus non currit.*) Ergo altera pars, videlicet, *Circulus habet quatuor Angulos*, erit necessario admittenda."

REDUCTIO   If from the assumption that all propositions of a certain set $\{P_1, ..., P_n\}$ are true, it can be concluded that, e.g., A and $\neg A$ should both be true, it follows that at least one of the P_i must be false!

After such an application of REDUCTIO, no sane logician would ever try to apply ECQ and argue that, since both A and $\neg A$ *have been shown to be true*, it follows that Socrates is a stone, and that Socrates is a lily; or that a square has four corners; etc. Of course, in a certain way it is legitimate to paraphrase ECQ à la Priest by saying that "from contradictory premises anything *follows*". But it appears much more adequate to paraphrase ECQ in the *subjunctive* (or *counterfactual*) mood: "If contradictory premises *would* be true, then anything else *would* be true as well". Caramuel was well aware of this important point for he introduced his proof of ECQ 1 by saying:

> But which bad things would occur in the world if, *per impossibile*, two contradictory propositions would be true together? Or, what if, *per impossibile*, one and the same proposition would be both true and false? I answer: Then in the whole world not a single truth would remain.[11]

To be sure, such "bad things" *would* happen, *if – per impossibile!* – two contradictory propositions like A and $\neg A$ *would* both be true. But don't worry: it is *impossible* that both A and $\neg A$ *are* true! Therefore, nobody has to be afraid that the set of true propositions "explodes" into the universal set. The latter scenario can be described, somehow paradoxically, by Caramuel's words that "not a single truth in the whole world would remain", for if every proposition would be true, then also all their negations would be true, and thus – at least in this sense – each proposition would be false.

---

[11]Cf. [9, p. 215]: "Sed quid in Mundo mali accideret, si per impossibile essent duae Contradictoriae simul verae? aut *Quid si per impossibile, una et eadem Propositio, vera & falsa esset simul?* Respondeo *Nullam in toto Mundo Veritatem mansuram.* Vel si dubites, da mihi has Propositiones, *Petrus currit,* & *Petrus non currit* (aut, si has nolis, alias quascunque Contradictorias) esse simul veras: & Ego tibi quidquid volueris demonstrabo."

3 Caramuel's Pentagon of Opposition

In his early logical works, Caramuel unconditionally subscribed to the correctness of the traditional square of opposition. Thus, in *Rationalis et realis philosophia* of 1642, he presented the following artful drawing:

Fig. 4: Traditional square of opposition

According to this diagram, the universal propositions
(A) 'Every man is white'
(E) 'No man is white'
are "contrary" to each other, i.e., they cannot both be true, but it is possible that both are false. In contrast, the particular propositions
(I) 'Some man is white'
(O) 'Some man isn't white'
are "subcontrary", i.e., they cannot both be false, but it is possible that both are true. According to the laws of *subalternation*, (A) entails (I), and (E) entails (O). Furthermore, (A) is contradictorily opposed, i.e., the *negation* of (O), and (E) is the negation of (I).

A few pages later, Caramuel presents the following variant of the Square with several paraphrases of the categorical forms:

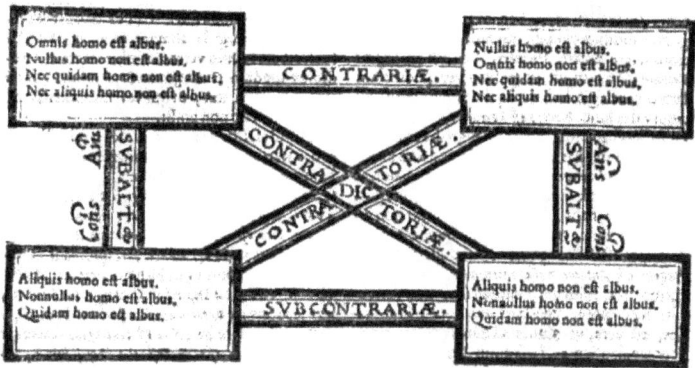

Fig. 5: Square of opposition with asymmetric paraphrases of the categorical forms

This schema contains some disturbing asymmetries. First, the universal propositions (A) and (E) are formulated in *four* different ways, while for the particular propositions (I) and (O) only *three* different formulations are given. Second, some paraphrases or (A) and (E) contain besides the normal negation '*non*' also the negation operator '*nec*'. Third, in all four forms, one paraphrase is formulated in terms of the quantifier expression '*quidam*' while another paraphrase is formulated in terms of '*aliquis*', but these two quantifiers are evidently synonymous!

The latter redundancy is best removed by simply *dropping* all paraphrases with 'quidam'. Furthermore, the second asymmetry is best removed by replacing all occurrences of 'nec' by 'non'. Thus, for the (A)-proposition there remain only three variants in terms of the informal quantifiers 'omnis', 'nullus', and 'aliquis':

(A) Omne S est P Every S is P
     Nullum S non est P No S isn't P
     Non aliquis S non est P Not some S isn't P.

Similarly, the three variants of the (E)-proposition are:

(E) Nullum S est P No S is P
     Omne S non est P Every S is not P
     Non aliquis S est P Not some S is P.

The last of the afore-mentioned asymmetries can now be removed by adding to the paraphrases of (I) and (O) based on 'aliquis' and 'nonnullus' another paraphrase based on 'non omnis':

201

(I)   Aliquis S est P   Some S is P
      Non nullus S est P   Not no S is P
      Non omne S non est P   Not every S is not P.

(O)   Aliquis S non est P   Some S isn't P
      Non nullus S non est P   Not no S isn't P
      Non omne S est P   Not every S is P.

These stratified versions of the categorical forms were formulated by Caramel twelve years after the publication of *Rationalis et realis Philosophia*. On p. 69 of *Theologia rationalis*, Caramuel first presented the square as in *Fig. 4* above, but a few pages later he added the following more elaborate diagram:

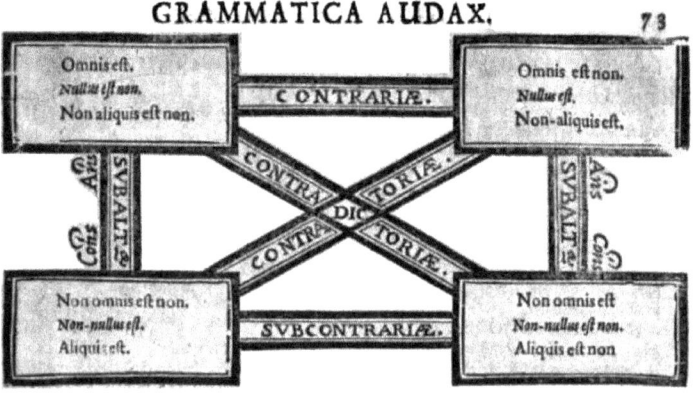

Fig. 6: Square of opposition with correct, symmetric paraphrases

Otherwise, however, also in this work Caramuel completely adopted the traditional theory of opposition, and it is not before the *Herculis Logici Labores Tres* of 1655, that he began to put this doctrine into doubt. In the first of the "Herculean tasks", the Square of Opposition is depicted in the usual way:

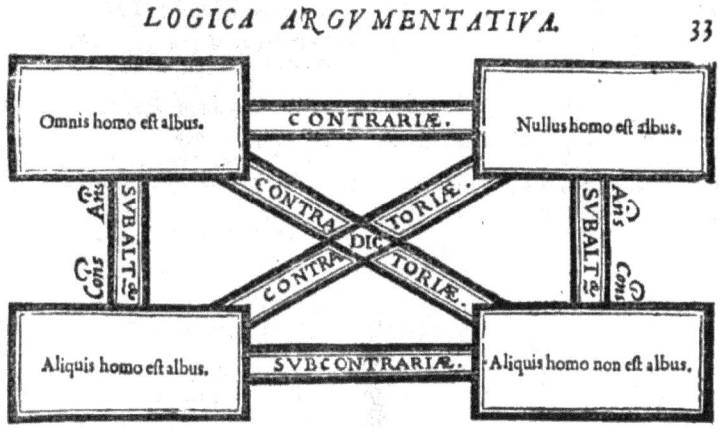

Fig. 7: Square of opposition with a self-critical remark at the left margin

But in the left upper margin Caramuel self-critically remarks that when he had written this passage, he had not yet observed the additional syncategorematic expression 'non omnis' which is not reducible to any of the four traditional ones.[12] Later on, within section 'De Terminis', he tries to explain this issue at some greater length:

> In the first place there occurs a distinction between categorematic and syncategorematic terms: we accept the definitions & to the syncategorematic terms which were posited by the ancients (namely, *Every*, *No*, *Some*) one has to add the definition of this one (*Not every*) which is treated here: and this is completed by the two subcontrary syncategorematic terms (namely, *Some* & *Some not*).[13]

In the subsequent section 'De Propositione', Caramuel defines all five

---

[12]The commentary on [6, p. 6] reads: "*Cum has dicebam, aut scribebam, nondum observaveram syncategorema non omnis nulli ex caeteris aequipollere & ideo loco particularem postulare.*" As a referee of this paper kindly pointed out, basically the same pentagon of opposition was considered already in the 14th century by Nicole Oresme (ca. 1320-1382). Cf. [10] where Caramuel's pentagons are briefly mentioned.

[13]Cf. [6, p. 28]: 'Occurrit primo loco Termini categorematici & syncategorematici distinctio: placent definitiones, & Syncategorematis a veteribus positis (videlicet, *Omnis, Nullus, Aliquis*) hoc ipsum (*Non omnis*) de qua agimus adjungi debere definit[io]: illudque duo syncategoremata subcontraria complecti, (nimirum *Aliquis*, & *Aliquis non*.)"

syncategorematic terms – and the corresponding categorical forms – explicitly as follows:

> The sign 'Every' constitutes the universal affirmative proposition; the sign 'No' the universal negative one; the sign 'Some' the particular affirmative proposition; the sign 'Some not' the particular negative one; & finally the sign 'Not every' constitutes the mixed particular proposition which both asserts and negates.[14]

Caramuel goes on to ask whether one and the same proposition might be both affirmative and negative. While other logicians use to answer this question in the negative, he argued that it should be answered in the affirmative because "this proposition 'Not every man is white', affirms and denies, for it is equivalent to 'Some men are white & some are not white'".[15]

Next Caramuel sets out to determine the logical relations between the four old and the new fifth categorical form. He first provides the following list of (very condensed) paraphrases:

(A) Every [S is P]. No [S is] not [P]. Not some [S is] not [P].
(E) No [S is P]. Every [S is] not [P]. Not some [S is P].
(I) Some [S is P]. Not no [S is P].
(O) Some [S is] not [P]. Not no [S is] not [P].
(Y) Not every [S is P]. Some [S is P] and some [S is] not [P].[16]

Then he recapitulates the usual definitions of the different kinds of opposition:

> There is an antipathy between propositions. Some are opposed only with regard to truth; they can be false together, but not true together,

---

[14]Cf. [6, p.28]: "Signum *Omnis*, constituit universalem asserentem: signum *Nullus*, negantem: signum *Aliquis*, particularem asserentem: signum *Aliquis non*, particularem negantem: & tandem signum *Non omnis*, particularem mixtam, simul asserentem et negantem."

[15]Cf. [6, p. 28]: "*An una et eadem esse possit simul affirmativa et negativa?* Negarunt hactenus: at hodie dicimus, hanc propositionem, *Non omnes homines sunt albi*, affirmare & negare: aequivalet enim huic, *Aliqui homines sunt albi, & aliqui non sunt albi*: quae partim affirmativa, & partim negativa est." Cf. also ibid.: "Et eodem modo possemus dicere hanc, *Non omnis homo est animal*, constare duabus propositionibus verae & falsae, adeoque totaliter non esse veram, sed partialiter."

Caramuel's Pentagon of Opposition

and they are called *contraries*. Some are opposed only with regard to falsity; they can be true together, but not false together, & they are called *subcontrary*. Some are opposed both with regard to truth and with regard to falsity; they cannot be true together, nor false together: & they are called *contradictories*.[17]

Then he asks (somewhat rhetorically) at which corner of the traditional square of opposition the new syncategorema 'Not every' has to be put; and he hastens to give the answer himself: "In no corner: but a special place has to be invented for it such that it takes care of the appropriate relation to the other terms".[18] In other words: The traditional *square* of opposition has to be extended to a *pentagon* in such a way that the former relations of contrariety, contradictoriness, and subalternation are preserved. As regards subalternation, Caramuel stresses that the traditional laws

SUB1   Every S is $P \Rightarrow$ Some S is P
SUB2   No S is $P \rightarrow$ Some S is not P

have to be supplemented by two further laws:

SUB3   Not every S is $P \Rightarrow$ Some S is P
SUB4   Not every S is $P \Rightarrow$ Some S is not P.[19]

As regards the relation of opposition between (Y) on the one hand end (A) and (E) on the other hand, Caramuel notes:

> I say first that the two propositions
> *Not every man is learned*
> *Every man is learned*
> and also the following two
> *Not every man is learned*

---

[17]Cf. [6, pp. 28-29]: "Est antipathia propositionum: q[uae] opponuntur veritate tantum, possunt esse simul falsae, non tamen simul verae, vocantur *Contraria*. Quae falsitate tantum, possunt esse simul verae, non tamen simul falsae, & vocantur *Subcontraria*. Quae veritate & falsitate simul, nec possunt esse simul verae, nec simul falsae: & vocantur *Contradictoria*."

[18]Cf. [6, p. 29]: "In quo ergo angulo ponetur syncategorema *Non omnis*? In nullo: sed sibi specialem locum inveniet, unde caeteros terminos relatione convenienti respiciat."

205

No man is learned

are contradictories: they can neither be together true, nor together false: at least in consensus and in the part, in which they are opposite. We have already seen that 'Not every etc.' is equivalent to two propositions and that it is partially opposed to each of the universal propositions (affirmative or negative) and that it partially entails them. Thus, one and the same proposition (what a miracle!) both contradicts and entails the same proposition, but, observe, with respect to different parts.[20]

In this work of 1655, Caramuel made no attempt to visualize these relations, but in *Moralis seu Politica logica* of 1680, he painted the following pentagon:[21]

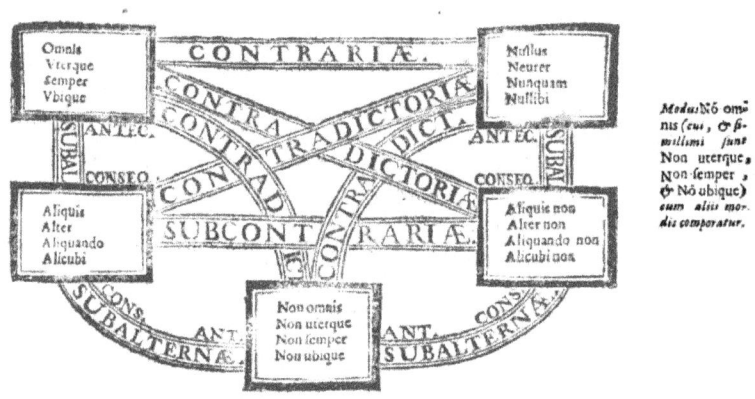

Fig. 8: *Caramuel's pentagon of opposition*

---

[20]Cf. [6, p. 29]: "Dico igitur primo has duas propositiones/ *Non omnis homo est doctus*:/ *Omnis homo est doctus*:/ et etiam istas duas/ *Non omnis homo est doctus*:/ *Nullus homo est doctus*:/ esse contradictorias: nec esse posse simul veras, nec simul falsas: saltem consensu et parte, qua sunt oppositae; q[uoniam] jam vidimus illam, *Non omnis* &c. duabus aequipollere: & universali alterutri (affirmativae aut negativae) partim opponi, partim subalternari. Ergo una & eadem propositio (mira res!) respectu ejusdem est contradictoria & subalterna: sed intellige, secundum diversas partes."

[21]The Scan is taken from [7, p. 25]. A similar pentagon may also be found in the part *"Logica Scripta"* of another work which Caramuel published in 1680: *Logica vocalis, scripta, mentalis, oblique, dialecticos locos dilucidans* ...; the diagram on p. 295 contains only the temporal "modalities" {Semper, Nunquam, Aliquando sic, Aliquando non, Non semper} and the spatial counterparts {Ubique, Nusquam, Alicubi sic, Alicubi non, Non ubique}.

On the one hand, the Pentagon of the "normal" expressions (A), (E), (I), (O), and (Y) is now also transferred to several other classes of "modalities" like 'Always', 'Never', 'Sometimes', 'Sometimes not', and 'Not always', etc. On the other hand, the relations of opposition between (A) and (Y) and between (E) and (Y) are still characterized as *'contradictories'*. But this, of course, is a *mistake*! If (Y) would be the negation both of (A) and of (E), it would follow that the negation of (A) would be identical to the negation of (E), but then (A) itself would be identical to (E)! Of course, in a certain way it is correct to say that proposition (Y) both *contradicts* (A) and *contradicts* (E). But (Y) is not therefore the literal *contradictory*, i.e., negation, of (A) and of (E); rather (Y) is *contrary* to both propositions![22]

4 Caramuel's Modal Pentagon of Opposition

In the later course of *Moralis seu Politica logica*, Caramuel investigates the (alethic) modal counterpart of the Pentagon. He points out that earlier logicians had always equated the expression 'contingent' with 'possible not'. But in the "*Herculean Tasks*" he had corrected this error by emphasizing that the concept of contingency contains both the possibility to be and the possibility not to be.[23] Therefore in analogy to the pentagon of *Fig. 8*, also the diagram for the opposition of the (alethic) modal operators has to contain a fifth corner:[24]

---

[22] As Demey noticed in fn. 8 of [10]: "Caramuel's pentagons [...] contain several mistakes. In particular, the relations from the Y-vertex to the A- and E-vertices are contrarieties [...], but Caramuel labelled them contradictories".

[23] Cf. [6, p. 50]: "Contingens dicitur, quod solet esse & non esse [...]. *Contingens* non aequipollet *Possibili*, ut crediderunt veteres: non aequipollet *Possibili non*, ut aliqui Iuniores malunt: sed utrumque complectitur, est enim contingens, quod & potest esse & non esse."

[24] Cf. [7, p. 34]: "Persuasum fuit ab initio Dialecticis, & multi etiamnum in hac opinione permanent, *Contingens* esse idem ac *Possibile non*. [...] Sed tamen, ut noster Hercules Logicus Labore III. evidenter ostendit, *Contigens* simul in suo conceptu includat esse posse et non esse; adeoque *Necessario* & *Impossibili* contradictorie opponatur, & subalternet, & per necessariam consequentiam inferat *Possibile sic*, & *Possibile non*. Ergo, sicut in Schemate, quo simplicium Propositionum Oppositionem repraesentavimus, quintam adhibuimus cellulam, in quam Syncategorema *Non omnis* poneretur; sic etiam hic quintam debuimus addere, ut Modum *Contingens* exponeremus."

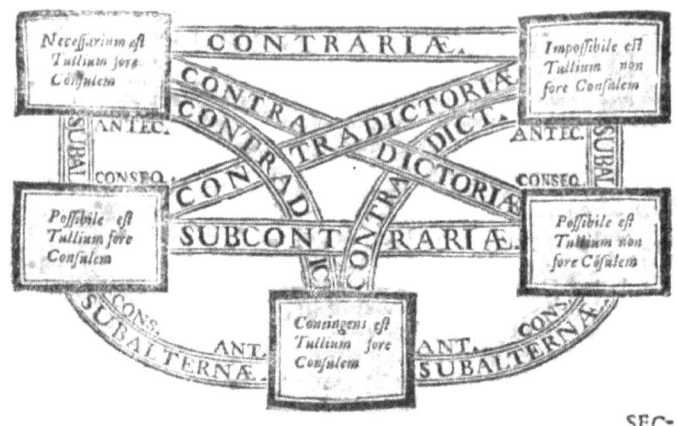

Fig. 9: Pentagon of (alethic) modal operators

Note that in this schema the (Y)-element, 'It is contingent that Tully becomes consul' is again erroneously classified as *contradictory* to the (A)-proposition 'Is is necessary that Tully becomes consul', and also *contradictory* to the (E)-proposition. But now Caramuel commits another small error when he formulates the (E)-proposition as 'It is impossible that Tully *does not* become consul' instead of 'It is impossible that Tully becomes consul'. This slip of the pen is, however, corrected a few pages later when Caramuel extends the modal pentagon to deontic modalities:

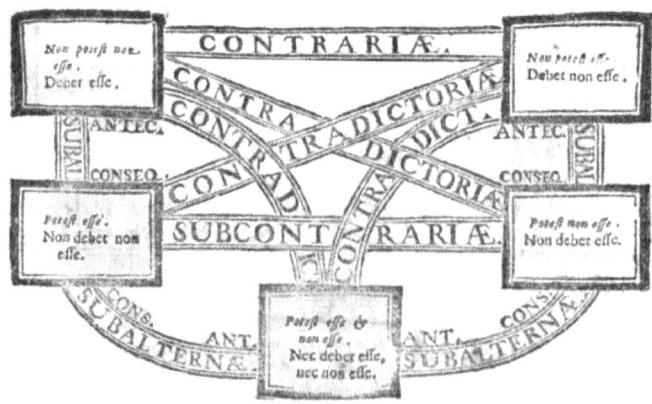

Fig. 10: Pentagon of alethic and deontic modalities

5 Completing Caramuel's pentagons

The incompleteness of Caramuel's modal pentagon is rather obvious. Since the (A) and the (E)-proposition are *contrary* but not *contradictory* to (Y), one simply has to add the *negation* of (Y) in a sixth corner so as to yield a *hexagon* of opposition. As in Blanché's [2], the sixth corner shall be symbolized by 'U'. In the case of *alethic* modalities, (U) amounts to the negation of 'potest esse & potest non esse', i.e., the negation of 'it is contingent that α', or in other words:

(U)$_{aleth}$ α is necessary *or* α is impossible.

Similarly, in the case of deontic modalities, the negation of 'Nec debet esse nec debet non esse', i.e., 'α neither has to be done nor α has to be omitted', can be defined thus:

(U)$_{deon}$ α is obligatory *or* α is forbidden

Finally, in the case of categorical propositions, the correct negation of Caramuel's expression 'Non omnis S est P' is not just 'Omnis S est P' but rather:

(U)$_{categ}$ Every S is P *or* No S is P.

References

[1] K. BERKA & S. SOUSEDÍK: 'On the relational logic of Juan Caramuel Lobkowitz' [in Czech], in *Acta Universitatis Carolinae: Philosophica et Historica, Studia Logica* 1972, pp. 9-16.

[2] R. BLANCHÉ: 'Sur l'opposition des concepts' in *Theoria* 19, 1953, pp. 89-130.

[3] R. G. BERRY (ed.): *Sextus Empiricus, Outlines of Pyrrhonism - With an English Translation..* Harvard University Press, Cambridge MA, 1933.

[4] J. CARAMUEL: *Rationalis et realis philosophia*, Leuwen (de Witte), also published as *Caramuelii Lobkowitzii Logica seu Philosophia Rationalis*, 1642.

[5] J. CARAMUEL: *Theologia rationalis sive in Auream Angelici Doctoris Summam Meditationes, Notae Observationes Liberales, Philosophicae, Scholasticae Tomi Duo*, Francofurti 1654 (Schönwetter).

[6] J. CARAMUEL: *Herculis logici Labori Tres*, Francofurti 1655 (Schönwetter).

[7] J. CARAMUEL: *Moralis seu Politica logica ...* Vigevano (Camillus Conrad), 1680.

[8] J. CARAMUEL: *Logica Vocalis, Mentalis et Obliqua*, Vigevano 1680. This book partly repeats passages from [5]; cf. [12, p. 646].

[9] J. CARAMUEL: *Leptotatos Latine Subtilissimus*, Vigevano (Camillus Conrad) 1681.

[10] L. DEMEY: 'Between Square and Hexagon in Oresme's *Livre du Ciel et du Monde*' in *History and Philosophy of Logic* 41, 2020, pp. 36-47.

[11] P. DVOŘÁK: *Juan Caramuel Lobkowitz: Selected Aspects of Formal and Applied Logic* [in Czech]. Oikumene, Prague, 2006.

[12] P. DVOŘÁK: 'Relational Logic of Juan Caramuel', in D. M. GABBAY & J. WOODS (ed.), *The Handbook of the History of Logic*, vol. 2: *Medieval and Renaissance Logic*. North Holland/Elsevier, Amsterdam, 2008, pp. 645-665.

[13] K. JACOBI (ed.): *Argumentationstheorie – Scholastische Forschungen zu den logischen und semantischen Regeln korrekten Folgerns*. Brill, Leiden, 1993.

[14] R. KILWARDBY: *Notule libri Priorum*, ed. by P. THOM & J. SCOTT. Oxford University Press, Oxford, 2015.

[15] W. LENZEN: 'Caramuel's Theory of Opposition' in *South American Journal of Logic 3, Special Issue Square of Opposition*, 2017, pp. 341-368, online: $http://www.sa-logic.org/start1.html$

[16] W. LENZEN: 'Kilwardby's 55th Lesson' in *Logic and Logical Philosophy* 29, 2020, pp. 485-504, online: $https://doi.org./10.12775/LLP.2020.010$

[17] W. LENZEN: 'The Third and Fourth Stoic Account of Conditionals' in M. BLICHA & I. SEDLAR (ed.), *The Logical Yearbook 2020*, College Publications, London, 2021, pp. 127-146.

[18] C. MARTIN: 'William's Machine' in *The Journal of Philosophy* 83, 1986, pp. 564-572.

[19] C. MARTIN: 'Embarrassing arguments and surprising conclusions in the development of theories of the conditional in the twelfth century' in J. JOLIVET & A. DE LIBERA (ed.) *Gilbert de Poitiers et ses contemporains: aux origines de la logica modernorum*, Bibliopolis, Napoli, 1987, pp. 377-400.

[20] A. NECKHAM: *De Naturis Rerum*, ed. by TH. WRIGHT. Longman, Roberts, & Green, London, 1863.

[21] G. PRIEST, K. TANAKA & Z. WEBER: 'Paraconsistent Logic' in E. N. ZALTA (ed.), *The Stanford Encyclopedia of Philosophy* (Summer 2018 Edition), online: https://plato.stanford.edu/archives/sum2018/entries/logic-paraconsistent/

[22] W. RISSE: *Die Logik der Neuzeit*, vol. 1. Frommann, Stuttgart, 1964.

[23] S. SOUSEDÍK: 'A Discrete Logic of Juan Caramuel Lobkowitz' [in Czech] in *Filosofický časopis* XVII, 1969, pp. 216-228.

Formal Logic Between Metaphysics and Purpose: Three Different Notions of 'Formal Fogic' in the Early 19th Century

VALENTIN PLUDER
University of Siegen
valentin.pluder@uni-siegen.de

1 Introduction

The meaning and purpose of any logic depend on the metaphysical framework in which it is embedded or for which it was conceived. If this is the case, the various logics that identify themselves as formal may mean something different by this attribute. This idea is supported by the current discussion of what it means when a logic is called 'formal'. In what follows, I will mention only three prominent positions as examples.

(i) Béziau [1] examines five different interpretations: 1. formal by opposition to matter, 2. formal by opposition to empirical science, 3. formal as in a formalized theory, 4. formal by using symbols rather than ordinary language or finally 5. formal by the use of mathematics. He concludes, however, that no concept of 'formal' is actually sufficient to describe the many possibilities and fields in which logic is currently used.

(ii) Dutilh Novaes [5] distinguishes eight different interpretations of 'formal' in formal logic, which can be divided in two main groups. The first one considers formal as pertaining to forms. In this case formal is the opposite of material, i.e. formal as 1. schematic, 2. indifferent to particulars, 3. topic-neutrality, 4. abstraction from intentional content or 5. desemantification. The second one considers formal as pertaining to the

213

strict application of rules, i.e. formal as 6. computable, 7. pertaining to regulative rules or 8. pertaining to constitutive rules. The opposite of this meaning of formal would be informal [5, p. 304].

(iii) In contrast, MacFarlane [13] works with a narrower selection of different interpretations because he asks more precisely in what sense formality can distinguish logic form other disciplines [13, p. 15]. He comes up with three different interpretations that are not mutually exclusive: 1. formal norms as 'constitutive of concept use as such', 2. formal as 'indifferent to the particular identities of different objects' or 3. formal as entirely abstract in regard of semantic content or matter [13, p. 51]. In addition, MacFarlane shows how the differences in the philosophies mainly of Kant and Frege lead to different ideas of 'formal'.

While it appears not to be clear what the meaning of 'formal' in formal logic actually is, it seems uncontroversial that there is a historical connection between the idea of a formal logic and the Kantian philosophy. On the one hand, Kant sees himself very clearly in the tradition of Aristotelian logic, on the other hand he is also the starting point and point of reference for discussions about logic and especially about formal logic at the beginning of the 19th century. This paper examines whether the assertion of a relationship between the meanings of 'formal' and the metaphysical framework in which a logic is placed is transferable to the debates immediately following Kant's logic. This historical approach to the subject suggests itself: Traditional logics usually provide explicit information about their purpose and the philosophical context in which they are placed. In connection with these logics, the terms 'logic' and 'system' are used in the sense of pre-Fregian logic.

Of course, many of the distinctions mentioned above cannot, or only with difficulty, be applied to the logics of the early 19th century. The possibility of mathematizing logic, for example, was hardly discussed at all and questions about desemantification and computability were not even on the table. Here it is, therefore, only examined whether a descriptive or a prescriptive concept is connected with 'formal logic', whether the forms of logic are constitutive for all thinking and what precise relationship there is between form and content in the particular formal logic. As examples of what formal logic meant in the first half of the 19th century, and to what extent the early 19th century logic can contribute to the current discussion, three writings on logic are used. These three chosen logics became very popular at the beginning of the 19th century. What these logics have in common is that they consider themselves 'formal'. However, they come from different traditions with different metaphysical assumptions and were written for different purposes.

First, there is the logic of August Twesten: *Grundriß der analytischen Logik* (*Outline of Analytical Logic*) from 1834 [20].[1] Twesten was primarily a theologian and as such a follower and successor of Friedrich Schleiermacher. In terms of his logic, however, he orients himself to Kant and his transcendental idealism. Second, there is the logic of Ernst Christian Reinhold: *Die Logik oder die allgemeine Denkformenlehre* (*Logic or the general theory of the forms of thinking*) from 1827 [18]. Ernst Christian is the son of the famous Carl Leonhard Reinhold, but unlike his father, he represents – at least in his logic – a metaphysical realism. Third, there is the logic of Moritz Wilhelm Drobisch: *Neue Darstellung der Logik nach ihren einfachsten Verhältnissen* (*New Presentation of Logic According to its Most Basic Relations*) from 1836 (first edition) [3]. Drobisch's logic is arguably the most influential of the three. Not least because Drobisch was a leading figure in Herbartianism, which was very popular in the 19th century.

2 Similarities in the three formal logics

'Formal logic' seems to be a Kantian coinage, although he does not use the expression very often. It is introduced in opposition to 'transcendental logic' and, more precisely, means a general and pure logic. Kant defines logic at large as 'the science of the rules of understanding' [11, p. B76] just to distinguish it into two branches: general logic and particular logic. The general logic 'contains the absolutely necessary rules of thinking, without which no use of the understanding takes place, and it therefore concerns these rules without regard to the difference of the objects to which it may be directed.' A particular logic 'contains the rules for correctly thinking about a certain kind of objects.'[11, p. B76]. The latter can be understood as methodology of special sciences. Whether

---

[1]The translation of the quotations from the texts available only in the original German was done by the author.

or not the transcendental logic is a particular logic is controversial [17, p. 301] [12, p. 265-266]. But it is certainly not general, because it deals with a 'determinate content, namely that of pure a priori cognition alone' [11, p. B170]. That means it 'has to do merely with the laws of the understanding and reason, but solely insofar as they are related to objects a priori.'[11, p. B81-82]. Against that the formal logic 'abstracts from all content of cognition (whether it be pure or empirical), and concerns itself merely with the form of thinking (of discursive cognition) in general' [11, p. B170]. In addition, formal logic is pure as it is not applied. Applied (but still general) logic would take into account the empirical-psychological conditions of the use of understanding [11, p. B78]. The term 'formal logic' as general and pure logic is coined within the framework of Kant's transcendental idealism [14, p. 44-56]. But it quickly spreads in philosophies that differ from Kant's position. This in turn leads to a different understanding of what 'formal logic' means in detail.

The three historical logics discussed here are similar in that they call themselves 'formal', that they all follow the classical structure of logic, and that they all want to be one system.

Just as Kant does not use the specific term 'formal logic' often, it cannot be found prominently in the books on logic by Twesten, Reinhold and Drobisch. However, it can be assumed that the three authors would not have disagreed if their logic were taken to be formal. First, because they all use the term 'form' or 'formal' to refer to their logic, and second, because they all regard their logic – justified or not – as pure and general. This is exactly how Kant characterizes formal logic.

In particular, Twesten states that his analytical logic is the 'science of the formal laws of thought' [20, p. 2]. In addition, Jäsche identifies an analytical logic with a formal logic [10, p. 10,11]. Since Twesten is largely committed to transcendental idealism, it is not surprising that he characterizes his own logic more or less in the same way like Kant [20, p. 11]: It is an a priori science [19, p. 9], which is general [20, p. 11] and pure [20, p. 10].

Reinhold's case is also very clear. The title of his book defines his logic as a theory of the general forms of thinking [allgemeine Denkformlehre]

[18, cf. p. III]. However, his logic is not only general, it is also pure in the sense of not empirical [18, p. 90] and in the sense of not applied, since there is no discussion about the empirical or psychological conditions of its use. Finally, Drobisch calls his logic 'formal philosophy' or 'mere formalism' [2, p. VI]. He deals with the general relationships of concepts [2, p. 3] and is devoid of any reflections on the psychological conditions of its use.

In addition, the formal character of the three logics goes hand in hand with the opposition to transcendental or metaphysical logics and is the result of abstraction. The three logics can therefore be described as 'formal'. But despite the same self-designation, the term 'formal' has different meanings for the different logics.

Initially, however, what catches the eye is the stereotypical structure of the logics. Like a majority of works on logic in the 19th century, the three logics roughly follow the traditional structure of Aristotle's *Organon*, which was established not by Aristotle himself but by editors in the first century BC, including Andronicus of Rhodos. Accordingly, Twesten, Reinhold and Drobisch deal at first with concepts (*Begriffe*) corresponding to the Aristotelian *Categories* (*categoriae*), then with judgments (*Urteile*) corresponding with Aristotle's *On Interpretation* (*de interpretatione*) before turning to arguments or inferences (*Schlüsse*) corresponding to Aristotle's *Prior Analytics* (*analytica priora*). In the 19th century, as well as in earlier periods, these three parts were normally followed by a fourth part on scientific methods, corresponding to the Aristotelian *Posterior Analytics* (*analytica posteriora*) and parts of the *Topics* (*topica*). The three logics differ with regard to the location of the laws of thought: Twesten places them at the beginning of his logic, with Reinhold they can be found after the inferences and Drobisch thematizes them in connection with the judgments.

At the time when the three authors wrote their texts on logic, what was generally known as 'logic' covered all these different subjects, and often even additional ones such as metalogical questions. A theory of deductive reasoning can be found – mainly in the form of syllogisms – in the third part of the *Organon*. Deductive reasoning in a strict sense is therefore only one element among others, and it was far from being the chief topic of scholarly interest at this time. The wide range of topics illustrates the urgency of a systematic connection. All three logics claim to be a system whereby system here means a whole in the sense of mutually supporting components. However, all three logics justify this in different ways.

3 Differences in the three formal logics

At first glance, the logics appear to be quite uniform. However, they differ significantly in their conception and treatment of some central points: Above all, they are based on different metaphysical assumptions. This results in different relationships to empirical experience, different objects of logic and different justifications for the systematic constitution of logic. Although logic is always understood as pure and general and thus as formal, there are also divergent understandings of what 'formal' should actually mean. Finally, the ideas about what the purpose of logic is diverge.

3.1 Twesten

Though he was no logician and maybe not even a philosopher in the narrower sense, the logic of August Twesten (1789-1876) became very popular in the first half of the 19th century. It shows quite clearly what 'logic' means and was meant to be for a great number of recipients of this kind of literature in this time. The metaphysical background of Twesten's logic is his interpretation of Kant's transcendental idealism, although he remarks that this is mainly the case for pragmatic reasons such as ease of understanding [19, p. XIX] [20, p. 7]. For Twesten that means that, there is a certain set of cognitive structures that exists independently of any experience or intuition and that can be expressed through concepts and laws of thought. Since this set describes the structure of a single and unified mind, this set forms a systematic unit a priori in itself. This is the reason why Twesten also regards logic as a system.

In addition, this cognitive structure forms the basis of all experiences, for experience is – somehow – the product of an unconscious synthesis of a priori forms with a posteriori content. These forms provide the forms for all intellectual activities without exception: Not only intuition and perception are always and directly interwoven with and through these structures, but this also applies to joint thinking [20, p. 3, 13]. However, all of these synthetizations do not fall within the domain of logic that is of interest to Twesten [19, p. 247]. As the name of his logic suggests, he is only (or primarily) interested in analytic logic.

In the eyes of Twesten, logic is a reflection of the a priori forms of thought. On the condition that all intellectual activities, including any kind of thinking, are shaped a priori in accordance with these structures and that logic at its core is the conscious reproduction of these unconscious forms a priori, it is clear that all thoughts must necessarily conform with the general forms of logic. This explains why Twesten believes that all people normally apply the laws of thought correctly unconsciously. This is a widespread thought in the 19th century, which can also be found, for example, in Schopenhauer [15, p. 130-132]. For the same reason, the concepts and laws of thought that reflect the a priori structure of all cognition do not need justification, but are simply evident [19, p. 15]. Moreover, the systematic constitution of the a priori forms reappears within the shape of logic. Twesten is convinced that – more or less – all of the logic can be derived from its central laws: the law of non-contradiction and the law of identity [20, p. 5-6, 11] (the principal of sufficient reason is banned from logic). Hence, Twesten's logic uses only these two laws – which are also assumed to be essentially one single law – to formally judge all thinking [20, p. 11]. In view of these metaphysical assumptions it is no surprise that Twesten points out that logic – as part of philosophy – deals with mere concepts and laws a priori independently of experience and intuition [20, p. 6-7]. Even though he notes that this premise may not be necessary or desirable in all respects, Twesten explains that it is indispensable for his logic [19, p. 48].

The purpose of his logic is to analyse thoughts solely in terms of their accordance with the laws of thought. Consequently, logic is not concerned with the question whether propositions have anything to do with empirical reality or with empirical knowledge at all [20, p. 5]. It is not devoted to any special kind of objects, as special objects ask for special concepts, while Twesten's analytical logic is strictly general [20, p. 4]. Finally, Twesten's logic is pure and not applied because it does not take into account the empirical conditions of its use [20, p. 10]. Like Kant Twesten characterizes the general and pure logic as 'formal': 'the formal laws of thought are subject of the common logic, that is called formal logic for that reason' [20, p. 4, cf. p. 2].

Twesten's analytical logic requires that something that can be analyzed be given upfront. So concepts and judgments are synthesized before Twesten's logic comes into play [19, p. 247], although he admits that sometimes the hand that put the finishing touches to them is guided by the analytical logic [19, p. 199]. For the most part, however, the laws of thought are used to clarify the relationships between concepts and judgments. Spoken language is the way in which concepts and judgments are given. They correspond to the thoughts that lay a sentences immediately below the words [20, p. 1, 13]: The forms of logic are used to analyse and reflect on the conscious and deliberate thinking only and '... this is the only kind of thinking common logic is concerned with.' [20, p. 4]

The forms of logic are generally valid because they are assumed to be the a priori conditions of all common thinking and once found, they are evident. But since there is no immediate awareness of them, they actually have to be found first. In Twesten's logic they are discovered through abstraction from their ordinary usage: The general concepts and the forms of logic appear when everything particular is removed, for thinking is the subsumption of the particular under the general by applying the laws of thought [20, p. 4, 15]. But of course that is not why the laws of thought and the general concepts are in force [20, p. 6-7].

The purpose of Twesten's analytical logic is primarily to describe the form of actual thinking. Operations that are carried out blindly according to the rules of logic are made conscious through the knowledge of logic. That does not mean that logic is never misused or that there never are logical fallacies, otherwise no logic would be required at all, but at large all thinking matches the forms of logic and corrections are more of an aesthetic nature. Once logic is known, it helps to arrange and organize thoughts, but still the 'effect of logic is more regulative than productive' [19, p. 11].

Formal Logic Between Metaphysics and Purpose

So what does the 'formal' in formal logic mean under these circumstances? First, it means that logic is constitutive of all thinking. Therefore, the logic has to be general. Second, the metaphysical background allows the 'formal' to be viewed as completely separate from any content. The forms have their own independent base. It is therefore possible to have pure forms. In terms of the philosophy on which Twesten's logic is based, this may seem plausible. However, it might be difficult to actually tell the difference between form and content, since form and content are always intertwined in any thinking. Twesten does not specify a criterion for differentiating between form and content. This problem becomes even more urgent with the logic of Reinhold.

3.2 Reinhold

Though Ernst Christian Reinhold (1793-1855) was – in the broader sense – a follower of the Kantian philosophy, the metaphysical framework in which Reinhold places his logic is that of realism. In contrast to Twesten he not only beliefs that the matter of perception exists independently of cognition. He also thinks that this independent reality is formed and organized by its own laws. The aim of cognition is to recognize and reproduce form and content of this 'outside' reality, which means that 'the ideal forms of our understanding are to match the real forms of the matter of cognition, like an image matches its model' [18, p. 55, cf. p. 50].

Unlike Twesten, Reinhold does not belief that form and content are originally separate from one other and have to be synthesized. Rather they are a unit from the start. Just like Twesten, Reinhold is also convinced that logic forms a system. The systematic shape of logic, however, is not due to the unity of the thinking mind but due to the unity of reality, which is supposed to be independent of the mind.

Based on these realistic assumptions, logic develops through abstraction and generalization from experience: Every content including all intuition is ignored because logic is the doctrine of the forms of thought like the title of Reinhold's work says already. Logic reflects 'only on the form of concepts and relations of concepts' [18, p. 90]. Despite the focus on the form Reinhold does not completely separate form and content [18,

p. 54]. For him, logic is always the form of a content. There is no form by itself.

Furthermore, the characterization as formal is not sufficient to define logic because the notion of form has to be differentiated internally [18, p. 54]: In addition to the general and pure form of logic there are the specific forms of cognition and imagination. It seems, therefore, that logic is determinated not only by its distinction from content, but also by distinction from other forms.

That implies in turn for the cognition of the independent reality that it not only has a genuine content but also a genuine form [18, p. 55]. Unfortunately, Reinhold does not explain if he is thinking of formal structures that mark phenomena as real in the sense of independent of the perceiving mind, or if he is only thinking of the forms of special logic that are opposed to the forms of general logic in the Kantian tradition.

The same goes for the imagination. Imagination feeds on the empirical cognition. It is tied to the recombination of perceptions once made [18, p. 56-57]. But since, in Reinhold's eyes, perceptions carry their own forms, the forms of imagination are not arbitrary either. They have to correspond to their empirical origin, which in turn represents the independent and formed reality. Again, Reinhold does not specify the specific forms of imagination. But it seems that he has forms in mind that restrict possible rearrangements of previous perceptions in ways that go beyond perimeters set by the purely formal concepts and laws of thought.

What imagination and cognition still have in common after abstraction from their specific forms are the formal laws and concepts of thought (or the pure and general logic): 'The same way in which our thoughts are bound both in knowledge and in imagination, in order to consciously represent objects in general, is the logical or general form of thought, as distinguished from the special forms of knowledge and imagination.' [18, p. 57-58]. But even after all specific forms have been abandoned Reinhold points out that the content or matter has always has to be kept in view [18, p. 54].

Logic expresses the forms and laws of thinking as such. The specific forms of cognition and imagination are not alternative to these general forms but additional. Moreover, Reinhold is not concerned with the empirical conditions of thought. In this respect too, his logic is pure. The ‚logical teachings of the forms of thought' could therefore be considered a formal logic, although Reinhold does not use precisely this term.

But is his logic really formal? The conception of formal logic as a result of the abstraction of content and special forms beginning from perception raises a number of questions. First of all, Reinhold gives no reason why these laws should be absolutely necessary and universally valid. Although he distinguishes the laws of logic from those of grammar, which sometimes seem to be random and arbitrary, he does not say why this difference exists. It is only postulated that being and thinking are structured by the same general laws and that these laws can be found through an abstract reflection on the individual act of thinking [18, p. 65-66].

This begs a second question. How can Reinhold be sure where the demarcation line between form and content lies? There is no criterion for precisely marking the point at which the abstraction process enters the realm of pure logic and, for example, leaves the field of grammar. On the contrary, he always points out that form and matter cannot be separated. The reason for this may simply be that his metaphysics does not allow a clear distinction between the two. One might wonder whether a logic that is ultimately based on the perception of an outside world that is independent from the mind and that unites form and content indistinguishably can be described as formal at all. At least it can't be formal in the same way that Twesten regards his logic as formal, because the metaphysical starting points of both philosophers are fundamentally different. Ultimately, this casts doubt on Reinhold's rejection of transcendental or metaphysical logic, particularly with regard to the alignment of logic with the conscious representation of objects in general.

The purpose of Reinhold's logic is quite similar to that of Twesten's logic. Like Twesten, Reinhold describes how thinking works in general. Since Reinhold's forms of thought have to mirror the forms of a reality that is supposed to be independently 'outside' of the thinking subject, there is a higher probability of misuse and fallacies compared to Twesten. In both cases, however, there is no fundamental difference between actual thinking and the laws of thinking.

So what does the 'formal' in formal logic mean under these circumstances? First, it does not mean that logic is directly constitutive for all thinking, as Twesten thinks it is. Since the mind reproduces the 'outside' world in form and contend, it could be said that both aspects are constitutive for the mind. But of course, that does not mean that the formal logic constitutes the 'outside' world. Second, Reinhold notes that 'formal' does not mean the complete absence of content or objects, but rather the general form that all content has after all special forms have been abandoned. Just as form and content are not separate but one in the outside world, the forms of thought cannot be separated from their content either. Even the most general forms are always the forms a content that is seen in the most general way.

3.3 Drobisch

'One of the most widely read logic treatises of the early nineteenth century was written by a *Herbartian*' [7, p. 10]: The philosophy of Moritz Wilhelm Drobisch (1802-1896) is not directly based on the Kantian philosophy but on the metaphysics of Johann Friedrich Herbart which can be considered as pre-Kantian. Herbart combines the conviction that all knowledge depends on empirical experience with a deep skepticism towards experience [9, p. 51-53, cf. 15-16,45]. He beliefs that all experience is marked by contradictions which are irresolvable within realm of experience. That renders empirical experience an elusive appearance in contrast to the actual reality, which is on the one hand independent from the subject, but on the other hand cannot be experienced but has to be understood through thinking. So in a certain way, Herbart is a realist like Reinhold only that for Herbart experience is the veil that has to be passed in order to access the reality behind it, whereas for Reinhold reality presents itself unobstructed in form and content in the experience.

The contradictions of the systematically inconsistent empirical experience must be resolved in order to gain knowledge of the actual reality. This is accomplished by distinction: Predicates that contradict each other are no longer assigned to just one entity, but each predicate is assigned to an entity on its own. The ultimate entities of reality only have one predicate. All contradictions are resolved in this way. However, these entities can no longer be perceived. They are beyond experience and need to be invented by the thinking mind [9, p. 35,37].

On the one hand, the fundamental distrust in empirical experience and the necessary rectification of its incoherence calls for an authority that is independent of experience. On the other hand, there is the necessary inclusion of experience due to the lack of other sources of knowledge. Drobisch's logic reflects these opposing conditions.

Since the point is to judge and correct empirical experience, no experience can influence the conceptualization of logic in the first place. The realm of logic is therefore exclusively that of conceptual thinking: neither perception nor intuition and also no metaphysical reflections are to be found in the field of logic, but rather in psychology and ontology [3, p. 9]. Drobisch states that philosophy and logic as a part of philosophy deal only with concepts. The purpose is 'to create knowledge exclusively from concepts' [3, p. 2]. Only after the abstraction of all contends does logic investigate the laws that rule these formal relations. Logic is therefore 'also called formal philosophy' [3, p. 3].

Again, logic is viewed as general and pure: general because it abstracts from all contend including any special methods and principles of understanding that would be related to special objects, pure because the empirical conditions of its application are of no interest [3, p. 89]. In contrast to Reinhold, all content is actually dropped. Hence logic 'is indeed nothing else but a mere formalism' [3, p. VI].

Like Twesten and Reinhold, Drobisch also regards logic as a system. Unlike the other two thinkers, he does not simply postulate the unity of all logical forms. He rather tries to generate the unity starting from the concept in a non-analytical – and in his eyes less dry – way [3, p. VII]. This is what he means when the title of his work promises an exposition of logic according to its simplest relations. The systematic presentation of logic gives the impression that all knowledge about logic is derived or deduced from these simplest relationships. In fact, however, Drobisch draws his knowledge of logic from experience.

Although it is necessary to perfectly separate logic from systematically corrupt experience, Drobisch, like Herbart, refuses to fall back on concepts that are a priori. In particular, the second edition of his logic, published in 1851, places special emphasis on this. Drobisch points out that knowledge about logic must have its origin in empirical experience [4, p. VI-V]. But the logician is not content with observing actual thinking. She rather '... uses the actual thinking as a fact to learn about the conditions of the lawful thinking' [3, p. 6]. Like the basic forms of geometry, the basic forms of logic are generated by abstraction from experience. And just as the ideal forms of geometry cannot be found within the realm of empirical experience, the ideal forms of thought cannot be averaged over the actual thinking [4, p. IV]. Logic describes thinking not as it is, but as it is supposed to be [3, p. 5-6].

Drobisch is aware that this is no justification for considering his logic as universally valid. In this respect, he restricts himself to a formal argument, but without elaborating on it: The justification of a logic must itself be subject to the laws of thought that are to be justified [4, p. 3-4]. In Drobisch's eyes, this is sufficient to ensure the truth of a certain logic, at least by and large.

The purpose of Drobisch's logic is to correct the thoughts that are misled by empirical experience. While the logic of Twesten and Reinhold is descriptive, Drobisch's logic can be characterized as prescriptive. Before logic is known, thinking is therefore a collection of statements that may correspond to experience but inevitably contradict each other.

So what does the 'formal' in formal logic mean under these circumstances? First, it does not mean that logic is constitutive of thought. In particular, thoughts that are close to intuition or empirical experience do not conform to logic. Second, logic must be independent of any empirical content in order to correct experiences. Since there is no content that does not come directly or indirectly from experience, Drobisch, like Twesten, regards 'formal' as completely indifferent to any content [4, p. IV]. Especially in view of his affinity to a mathematization of logic, Drobisch seems to be on the verge of desemantification. But that line is actually not crossed by him [3, p. VIII].

4 Conclusion

It has been shown that the meaning of the term 'formal' depends on the metaphysical assumptions and purpose of a particular logic. Three variants of logics that consider themselfs 'formal' were presented, some of which overlapped and some of which differed from each other:

Twesten and Reinhold represent a descriptive concept of formal logic. However, they advocate this concept for different reasons. For Twesten, the logical forms are constitutive for all thinking. They are therefore also constitutive for all reality-related thoughts. For Reinhold, the form of reality itself, regardless of the thinking subject, corresponds to the form of logic. Form and content are always linked in reality, which is independent of the thinking subject. With Twesten, form and content are initially separated. Experience is constituted only with recourse to forms that are independent of any experience. In contrast, with Reinhold, form and content are experienced as one unit. Therefore, form and content cannot be completely separated even in thought. In formal logic, too, the forms always remain united with a general content. Twesten and Drobisch, on the other hand, see the forms of logic as completely separate from any content. In contrast to Twesten, however, with Drobisch these forms do not constitute all thinking a priori. Rather, the forms of logic correct the thinking that arises directly from experience. The logical forms, which are completely independent of experience, are therefore prescriptive in Drobisch.

The question dealt with here aimed only at the differences in the meaning and purpose of formal logics depending on the metaphysical framework. It would be interesting to examine whether the metaphysical framework also influences the determinations within logic: the possible forms of concepts, judgments, inferences etc. Here, too, it would seem obvious to separate the form of logic from the purpose and meaning of logic. Then it could turn out that these uninterpreted forms actually coincide – at least at this stage in the history of logic. But here a problem arises again, which was particularly evident with Reinhold, but also with Twesten and Drobisch: The criterion for distinguishing between form and content would have to be specified. In particular, this criterion must of course be independent of whether the apparently purely formal inner-logical determinations of the individual logics coincide or not.

Already at the beginning of the 19th century, not long after Kant coins this term, there is a plurality in the meaning of 'formal logic'. This is obvious with the logics of this time, because it was part of the canon of logic to provide information about the metaphysical presuppositions. The differences in regard of these metaphysical presuppositions lead to different meanings of 'formal', as can be seen from the logics of Twesten, Reinhold and Drobisch. At the same time, it is noticeable that the 'formalism', i.e. the basic assumptions and rules of these traditional logics, is generally quite stereotypical. This can mean two things: Either the 'formalism' is independent of the meaning of 'formal' or at least some logics are inconsistent in this regard because they simply graft the traditional logic onto their metaphysical beliefs. The same question about the systematic connection between formalism and the meaning of 'formal' can of course also be asked in connection with the current debate. If such a connection exists, it would have to be clarified in what sense such logics can still be described as 'formal' or whether 'formal' describes a certain connection or a certain group of connections. Béziau, Novaes, and MacFarlane claim after all that there are a number of meaningful meanings of 'formal'. This variety of meanings presumably results systematically from the difficulty of formally clarifying the relationship between form and meaning. The formal logic may therefore not be formal enough to be defined as formal in a uniform way: 'their mistake is not that they are only form, but that they lack form' [8, p. 239]. This dictum can also be found at the beginning of the 19th century. However, it comes from a side that is far from being suspected of wanting to pursue formal logic [16].

References

[1] J.-Y. Béziau, 'What is 'formal logic'?' in *Proceedings of the XXII. World Congress of Philosophy*, 2008,13, pp.9-22.

[2] M. W. Drobisch, *De calculo logico: Programma quo ad audiendam orationem muneris professoris matheseos publici ordinarii.* C. P. Melzer, Lipsiae, 1827.

[3] M. W. Drobisch, *Neue Darstellung der Logik nach ihren einfachsten Verhältnissen. Nebst einem logisch-mathematischen Anhange.* L. Voss, Leipzig, 1836.

[4] M. W. Drobisch, *Neue Darstellung der Logik nach ihren einfachsten Verhältnissen. Nebst einem logisch-mathematischen Anhange.* 2. Auflage L. Voss, Leipzig, 1851.

[5] C. Dutilh Novaes, 'The Different Ways in which Logic is (said to be) Formal' in *History and Philosophy of Logic*, 2011, 32:4, pp. 303-332.

[6] C. Dutilh Novaes, *Formal Languages in Logic. A Philosophical and Cognitive Analysis*. University Press, Cambridge, 2012.

[7] J. Ferreirós, 'Traditional Logic and the Early History of Sets' in *Archive for History of Exact Sciences*, 1996, 50:1 pp. 5-71.

[8] G. W. F. Hegel, *Vorlesungen über die Geschichte der Philosophie II*. In Werke 19, Suhrkamp, Frankfurt a. M., 1986.

[9] J. F. Herbart, *Allgemeine Metaphysik nebst den Anhängen der philosophieschen Naturlehre. Zweiter, systematischer Teil.* (1829). In Sämtliche Werke VIII, 2. Reprint, Scientia Verlag, Aalen, 1989.

[10] G. B. Jäsche, *Immanuel Kants Logik.* F. Nicolovius, Königsberg, 1800.

[11] I. Kant, *Critique of pure Reason*. Translated and edited by P. Guyer and A. W. Wood. University Press, Cambridge, 1998.

[12] S. Käufer, 'Hegel to Frege: Concepts and Conceptual Content in Nineteenth-Century Logic' in *History of Philosophy Quarterly*, 2005, 22:3, pp. 259-280.

[13] J. McFarlane, *What does it mean to say that logic is formal?* PhD Dissertation, University of Pittsburgh, 2000: johnmacfarlane.net/dissertation.pdf.

[14] J. McFarlane, 'Frege, Kant and the Logic in Logicism' in *The Philosophical Review*, 2002, 111:1, pp. 25-65.

[15] V. Pluder, 'Schopenhauer's Logic in its Historical Context' in J. Lemanski (ed.) *Language, Logic and Mathematics in Schopenhauer*. Birkhäuser, Cham, 2020, pp. 129-143.

[16] V. Pluder, 'The Limits of the Square: Hegel's Opposition to Diagrams in its Historical Context' in I. Vandoulakis, J.-Y. Béziau (eds.) *The Exoteric Square of Opposition*. Birkhäuser, Cham, 2022.

[17] R. Pozzo, 'Kant within the Tradition of Modern Logic: The Role of the 'Introduction: Idea of a Transcendental Logic" in *The Review of Metaphysics*, 1998, 52:2, pp. 295-310.

[18] E. C. J. Reinhold, *Die Logik oder die allgemeine Denkformenlehre*. Crötersche Buchhandlung, Jena, 1827.

[19] A. D. C. Twesten, *Die Logik, insbesondere die Analytik*. Königliches Taubstummen-Institut, Schleswig, 1825.

[20] A. D. C. Twesten, *Grundriß der analytischen Logik*. Schwers Wittwe, Kiel, 1834.

Non-Denoting Terms and "Imaginary" Numbers: Modern Interpretations of Husserl's 1901 Double Lecture

Víctor Aranda
University of Salamanca
vic.aranda@usal.es

Abstract

This paper presents modern interpretations of Husserl's solution to the so-called "problem of imaginary numbers". The starting point of these interpretations are three different points of view on non-denoting terms: the many-sorted, the "nonexistent values" perspective and the "partial" approach. I will maintain that the last one is the most appropriate one for a reassessment of his Double Lecture, delivered in 1901. My proposal, that incorporates the existence predicate of free logics, gives an account of Husserl's "existence axioms", vindicating its essential role in the transition through the Imaginary.

1 Introduction

In the last two decades, much attention has been devoted to Husserl's 1901 Double Lecture (see, among others, Da Silva [5, 6], Hartimo [9] and Centrone [3]). A consensus has emerged that his Double Lecture, known today as *Doppelvortrag*, is an investigation of the conditions under which the enlargement of the system of natural numbers is permitted. This enlargement must be effected by ensuring that the rules of operation for the new[1] numbers do not contradict the laws that were valid for

---
I thank the two anonymous referees for their helpful comments on this manuscript.
[1] These new numbers are *imaginary*. Husserl used the term "imaginary" in a broad sense which includes every non-natural number [11, p. 412].

the calculations with natural numbers. According to Husserl, if the axiom system for the naturals is *relatively definite* and the theory of the enlarged system is *relatively* – or *absolutely* – *definite*, then this process will not lead to contradictions. It is important to remark, however, that he defined "relative definiteness" in the following terms:

> An axiom system is *relatively definite* if, for its domain of existence it indeed admits of no additional axioms, but it does admit that for a broader domain the same, and then of course also new, axioms are valid. New axioms, since the old axioms alone in fact determine only the old domain [...].
>
> The new axiom system defines a domain which includes the old domain and, consequently, in a certain way, also has all the old axioms in itself. [11, p. 426]

From a modern perspective, it is easy to check that the system of, for instance, the integers, does not satisfy all the axioms of the natural numbers. Let δ be $\forall x(sx \neq 0)$, that is, Peano's third axiom, which is saying that 0 is not the successor of a number. Obviously, δ is not true in the system of the integers and, in fact, it is false in every numerical system containing -1. Therefore, Husserl was wrong when he argued that *all* the axioms of the starting "domain" should be valid for the enlarged system. The question is what sentences are really preserved under successive enlargements of the system of natural numbers (without contradiction).

Da Silva [5, p. 422] has provided a more nuanced picture of Husserl's solution. Let $\mathcal{A}$ be a mathematical structure (for instance, $\langle \mathbb{N}, 0, +, \cdot \rangle$), and $Th(\mathcal{A})$ be the set of first-order[2] sentences φ of its language such that $\models_\mathcal{A} \varphi$. A sentence $\varphi \in Th(\mathcal{A})$ is preserved under successive enlargements of $\mathcal{A}$ if φ refers "exclusively to the elements of the domain of $\mathcal{A}$" [5, p. 422]. The sentence $\neg \exists x(x = \sqrt{-1})$ cannot be preserved under the transition from the real to the complex numbers, for the reason that

---

[2] As Centrone [3, p. 167] has pointed out, the paradigm of logic was not restricted to first-order languages at the beginning of the twentieth century. However, Da Silva assumed first-order logic as the most adequate logic for formalizing Husserl's intuitions [5, p. 423]. In modern logic, the set of first-order sentences φ of an appropriate language such that $\models_\mathcal{A} \varphi$ is called "theory of $\mathcal{A}$" [4, p. 37].

$\sqrt{-1}$ refers to a number, i, which is not in $\mathbb{R}$. Husserl himself offered the same explanation to the fact that the equation $x^2 = -a$ has no solution in the system of the integers:

> Let us consider, for example, the axiom system of the whole numbers, positive and negative. Then, $x^2 = -a$, $x = \pm\sqrt{-a}$ certainly has a sense. For square is defined, and $-a$, and = also. But "in the field" there exists no $\sqrt{-a}$. The equation is false in the field, since such an equation cannot hold at all in the field. [11, pp. 438-39]

Terms whose reference is outside the universe of discourse are usually considered *non-denoting*. For this reason, it is possible to reformulate Da Silva's interpretation as follows: A sentence $\varphi \in Th(\mathcal{A})$ is preserved under successive enlargements of $\mathcal{A}$ if φ does not contain any non-denoting expression. Since $\neg \exists x(x = \sqrt{-1})$ contains the non-denoting term $\sqrt{-1}$, the truth of $\exists x(x = \sqrt{-1})$ in $\mathcal{B}$ (the field of the complexes) does not contradict its falsity in $\mathcal{A}$ (any other system of numbers). In modern logic, there are many approaches for dealing with non-denoting terms [8, pp. 1271-75]. The aim of my paper is to find the most appropriate one for a reassessment of Husserl's answer to the "problem of imaginary numbers[3]".

Centrone [3, pp. 176-77] defends that the main difficulty of Da Silva's reading lies in the formal delimitation of the sentences which are preserved under successive enlargements of $\mathcal{A}$. In other words, the difficulty is how to *restrict* the sentences of the language of $Th(\mathcal{A})$ to those that refer exclusively to the elements of the domain **A** of $\mathcal{A}$. The set of sentences that $Th(\mathcal{A})$ can prove or disprove is called "apophantic domain" of $Th(\mathcal{A})$ [5, p. 427]. In this paper, I present the many-sorted and the "nonexistent values" approaches to non-denoting terms as possible tools for restricting quantification. I will assess the plausibility of both approaches and argue that the latter fits with Da Silva's [6] reading of Husserl's 1901 *Doppelvortrag*. Then, I introduce the "partial" perspective on non-denoting terms, which I consider more natural. This point of

---

[3] The problem of the conditions under which the enlargement of the system of natural numbers is permitted is called "problem of imaginary numbers" in the literature and "transition through the Imaginary" by Husserl.

view, together with the resources of free logic, provides a new interpretation of Husserl's solution based on his notion of "existence axioms". In the present article, and following Da Silva [6, p. 1926], I prefer to focus on *conceptual* analyses, rather than discussing textual and contextual aspects[4].

2 Many-sorted vs. "Non existent values"

2.1 The Many-sorted interpretation

The square root function is total iff it is defined as a function $\mathbf{f}: \mathbb{R} \longrightarrow \mathbb{C}$, but it is not total if its range is $\mathbb{R}$. From a many-sorted perspective, the sentence $\exists x(x = \sqrt{-1})$ refers to an element *outside* the universe of discourse because the appropriate "sorts" are not available. When the sort of the complex numbers is available, the term $\sqrt{-1}$ does have a denotation.

Let σ and ρ be the sort symbols for the real and complex numbers, respectively. By means of these symbols, it is possible to distinguish between different universes[5] of discourse in the language of $Th(\mathcal{B})$ (the enlarged system of numbers). Therefore, the sentence $\neg \exists x_\sigma(x_\sigma = \sqrt{-1})$ expresses the fact that there is no *real* square root of -1, while $\exists x_\rho(x_\rho = \sqrt{-1})$ is saying that there exists at least a *complex* square root of -1 (of course, $\exists_{=1} x_\rho(x_\rho = \sqrt{-1})$ establishes that the complex square root of -1 is unique). The possibility of distinguishing between different universes of discourse is compatible with the following passage by Husserl:

> Object domain of A (defined by means of A). Object domain of A_w (defined by means of A_w) [...].
>
> Imaginary objects = objects which do not occur in A, are not defined there, are not established by means of the axioms and

---

[4]"Of course, text and context are important and will be taken into consideration, but sometimes they can mislead rather than lead. A good deal of attention must be given first to the problems Husserl was facing and what they involved conceptually" [6, p. 1926].

[5]An interpretation for a many-sorted language with two sort symbols must guarantee that every term denotes an object in the appropriate universe. It follows that a non-denoting term in the sort of the real numbers can have a denotation in the sort of the complex numbers.

existential definitions of A, so that, therefore, if we regard A as the axiom system of a domain which has no other axioms –and thus also no other objects- those objects are in fact "impossible". [11, p. 433]

Obviously, the enlarged system of numbers $\mathcal{B}$ will have two universes $\mathbf{B_1}$ and $\mathbf{B_2}$ corresponding to the real and complex numbers. This many-sorted structure could be converted into a one-sorted mathematical system and, similarly, the many-sorted language could be easily converted into a one-sorted one. It is well-known[6] that the reduction of a many-sorted logic to a one-sorted formal system is performed on two levels (semantically and syntactically). I will defend that the syntactical translation of many-sorted formulas into one-sorted formulas is useful to *restrict* quantification to the elements of $\mathbf{A}$ (in other words, to the objects which are not "imaginary" from the point of view of $Th(\mathcal{A})$).

To translate many-sorted formulas into one-sorted formulas, consider a language whose variables range over a unique domain of quantification. This language includes two unary predicates R_σ and R_ρ, instead of the sort symbols σ and ρ. Thus, the universally quantified sentences have the form

$$\forall x_1, ..., x_n(R_\sigma(x_1, ..., x_n) \rightarrow \Psi(x_1, ..., x_n)) \text{ or}$$
$$\forall x_1, ..., x_n(R_\rho(x_1, ..., x_n) \rightarrow \Psi(x_1, ..., x_n));$$

the existentially quantified sentences

$$\exists x_1, ..., x_n(R_\sigma(x_1, ..., x_n) \wedge \Psi(x_1, ..., x_n)) \text{ or}$$
$$\exists x_1, ..., x_n(R_\rho(x_1, ..., x_n) \wedge \Psi(x_1, ..., x_n)).$$

Therefore, the sentences $\neg \exists x_\sigma(x_\sigma = \sqrt{-1})$ and $\exists x_\rho(x_\rho = \sqrt{-1})$ are translated into $\neg \exists x(R_\sigma(x) \wedge x = \sqrt{-1})$ and $\exists x(R_\rho(x) \wedge x = \sqrt{-1})$, respectively.

On the semantical level, the reduction is performed by an unification of domains. If $\mathbf{B_1} = \mathbb{R}$ and $\mathbf{B_2} = \mathbb{C}$, then it is clear that $\mathbf{B_1} \cup \mathbf{B_2} = \mathbb{C}$ (the universes of discourse do not have to be disjoint). Then, the operations defined on $\mathbf{B_1}$ and $\mathbf{B_2}$ are extended to the unified domain. Obviously, the square root function becomes *total* as soon as the values of the square

---

[6]See Manzano [12, pp. 221-22].

roots for negative numbers are available. These values were "imaginary numbers" in $\mathbf{B_1}$. It should be pointed out that the sentences which refer exlusively to the elements of $\mathbf{B_1}$ can be expressed in the one-sorted language and, in fact, its truth is preserved in the one-sorted structure. The sentence $\exists x_\sigma(x_\sigma = \sqrt{2})$ is true in the many-sorted structure, just like the sentence $\exists x(R_\sigma(x) \wedge x = \sqrt{2})$ will be true in the one-sorted one. The truth of $\exists x(R_\sigma(x) \wedge x = \sqrt{2})$ is independent from the axioms of the complex numbers, since it does not presuppose the validity of any "supplementary" axiom besides those which were true for the reals:

> The situation now is that we derive from A_w an assertion which refers purely to the objects of A. That presupposes: We see on the face of an assertion, or can prove at any time, that it has a "sense" purely for the objects of the narrower domain, i.e., that it, if it is true, presupposes the validity of no concept (the being of no object) which owes its validity only to the supplementary axioms. [11, p. 433]

Hence, the unary predicates R_σ and R_ρ allow the relativization of quantifiers to the real and complex numbers, and sentences of the form

$$\forall x_1, ..., x_n(R_\sigma(x_1, ..., x_n) \rightarrow \Psi(x_1, ..., x_n)) \text{ and}$$

$$\exists x_1, ..., x_n(R_\sigma(x_1, ..., x_n) \wedge \Psi(x_1, ..., x_n))$$

are interpreted as ranging over $\mathbb{R}$. These sentences constitute the apophantic domain of the theory of the reals (i.e. the collection of sentences that this theory can prove or disprove), what means that they are *all* preserved under successive enlargements of the model. Notice that there is no contradiction between the fact that $\neg \exists x(R_\sigma(x) \wedge x = \sqrt{-1})$ follows from the theory for the unified domain and the provability of $\exists x(R_\rho(x) \wedge x = \sqrt{-1})$ from the same axiom system. The reason is that, due to the relativization of quantifiers, these are different formulas.

With the relativization of quantifiers at hand, the notion of "relative definiteness" can be revisited. The sentence $\exists x(R_\sigma(x) \wedge x = \sqrt{-1})$ is false in the unified domain, because there is no $\mathbf{x}$ in $\mathbb{C}$ satisfying this formula when the scope of the existential quantifier is restricted to $\mathbb{R}$. A theory whose apophantic domain contains sentences that are false for this reason is called *relatively* definite. By contrast, in an *absolutely* definite axiom system, "no proposition meaningful in virtue of the axioms becomes false because of the fact that it has recourse to operational formations which are not defined" [11, p. 451].

Nevertheless, some objections could be raised against the many-sorted approach to non-denoting terms. A defect of this perspective, which has been already observed by Farmer [8, p. 1273], is the proliferation of sort symbols. This defect is specially serious in the context of the problem of imaginary numbers. Clearly, if the successive enlargements of the system of the naturals conclude with the complex numbers, then the many-sorted language must include *five* sort symbols. Similarly, the one-sorted language will have five unary predicates to restrict quantification to different domains ($\mathbb{N}$, $\mathbb{Z}$, $\mathbb{Q}$, $\mathbb{R}$ and $\mathbb{C}$).

Another important difficulty lies in the process of enlarging a system of numbers. It is doubtful whether the transition from the naturals to the integers, rationals and so on is carried out by adding to the language various sort symbols (or, equivalently, the corresponding unary predicates). Furthermore, there is no evidence in Husserl's *Doppelvortrag* supporting the syntactic translation of formulas of $Th(\mathcal{A})$ to formulas of $Th(\mathcal{B})$.

2.2 The "Nonexistent values" interpretation

According to Farmer, the attitude towards non-denoting terms depends on the philosophical point of view adopted[7]. "In short, the mathematician does not worry about what nondenoting terms mean; the logician does" [8, p. 1271]. Thus, some philosophers (Meinong) would argue that the denotation of these terms are *nonexistent* objects. I will show that the "nonexistent" approach to non-denoting terms fits well with the interpretation of the *Doppelvortrag* defended by Da Silva [6].

Unlike the many-sorted approach, a nonexistent perspective does not demand the addition of sort symbols to the language of the theory. Instead, it will be necessary to include a unique unary predicate R,

---

[7]"Most of the work done on logic with non-denoting terms has been motivated by questions concerning definite descriptions [...] The perspective of the (philosophical) logician who is interested in non-denoting definite descriptions is different from that of the mathematician who would like to reason about partial functions" [8, p. 1271].

what is also established in Da Silva's proposal: "let's introduce a unary predicate symbol R in $\mathscr{L}$" [6, p. 1931]. Of course this predicate is intended to *restrict* quantification, but now the quantifiers of $\mathscr{L}$ do not range over different sorts that could be eventually unified. On the contrary, the universe of a mathematical structure $\mathcal{A}$ will be **A** *together with* a set of nonexistent values.

This perspective on non-denoting terms should be clearly distinguished from the so-called "error value" approach (which was common currency in computer science). Error values are considered to be "first class objects" [8, p. 1273], that is, the operations defined on a domain may produce the error value $*$ as its outputs. Thus, the range of the square root function **f** is $\mathbb{R} \cup \{*\}$ –when the complex numbers are not available. When $\mathbf{f}(-1) = *$, it is evident that the denotation of $\sqrt{-1}$ will be $*$ (the inclusion of this value allows to convert a partial function to a total mapping). The nonexistent values, however, are considered to be "second class objects", since quantifiers do not range over them. The purpose of the unary predicate is, precisely, to guarantee that quantification is restricted to existent objects.

From this point of view, the universe of a mathematical structure $\mathcal{A}$ will be the union of a "inner" domain of existent values and an "outer" domain of nonexistent objects [8, p. 1273]. A denoting term of the language of $Th(\mathcal{A})$ has its reference in the inner domain, while a non-denoting expression "denotes" an element of the outer domain. "Quantification is only over existent values [...] Free variables range over values in both the inner and outer domains" [8, p. 1273]. Such a restriction on the scope of the quantifiers is carried out by imposing that universally and existentially quantified are of the form

$$\forall x_1, ..., x_n (R(x_1, ..., x_n) \to \Psi(x_1, ..., x_n)) \text{ and}$$

$$\exists x_1, ..., x_n (R(x_1, ..., x_n) \land \Psi(x_1, ..., x_n))$$

In order to explain the semantics of the unary predicate R, Da Silva asks the reader to consider a subset **D** of the domain **A**. "We say that φ_R refers to **D** in **A** if $R^{\mathcal{A}} = \mathbf{D}$" [6, p. 1931]. In other words, an universally or an existentially quantified sentence including the predicate R refers exclusively to the elements of **D** iff the interpretation of R in $\mathcal{A}$ is the set **D**. It is easy to see that the "inner domain" can be regarded as a subset of the set that results from the union of the "outer" and the inner domains. Thus, it is always possible to restrict quantification to the set of existent objects by means of the unary predicate R. Now, suppose that the domain of $\mathcal{A}$ is the union of $\mathbb{R}$ and a collection of nonexistent objects. Clearly, the sentence $\exists x(R(x) \wedge x = \sqrt{-1})$ is false in $\mathcal{A}$ when $R^{\mathcal{A}}$ is fixed to be $\mathbb{R}$. If the inner domain is enlarged with the complex numbers, $\exists x(R(x) \wedge x = \sqrt{-1})$ is true in $\mathcal{B}$, when $R^{\mathcal{B}}$ is fixed to be $\mathbb{C}$.

The provability of $\exists x(R(x) \wedge x = \sqrt{-1})$ from $Th(\mathcal{B})$ does not contradict the fact that $\neg \exists x(R(x) \wedge x = \sqrt{-1})$ follows from $Th(\mathcal{A})$, for the reason that the meaning of R in $\mathcal{A}$ (the reals) is different from its interpretation in $\mathcal{B}$ (the complexes). In other words, the inner domains of $\mathcal{A}$ and $\mathcal{B}$ do not coincide. I think that the "nonexistent" approach also gives a plausible account of relative and absolute definiteness. In my view, a relatively definite theory could be defined as an axiom system whose intended model has an outer domain. By contrast, an absolutely definite set of sentences could be regarded as a theory whose model does not have an outer domain (that is, all the elements of the domain of $\mathcal{A}$ exist). In Da Silva's terminology, $R^{\mathcal{A}} = \mathbf{A}$, what means that the interpretation of R in $\mathcal{A}$ is *the domain* $\mathbf{A}$. The sentence $\exists x(R(x) \wedge x = \sqrt{-1})$ of the language of an absolutely definite theory is, hence, equivalent to $\exists x(x = \sqrt{-1})$. "Obviously, if R refers to $\mathbf{A}$, then φ and φ_R are either both true or both false in $\mathcal{A}$" [6, p. 1932].

As a result, the transition through the Imaginary might be a process of formation of new existent objects that were "impossible" (i.e. nonexistent) without considering additional axioms. It follows that the model of the (relatively definite) theory for the naturals should have the "biggest" outer domain, which is reduced under successive enlargements of the inner domain (by means of the negative, rational numbers, etc.). Finally, the model of the (absolutely definite) theory for the complex numbers should have the "smallest" outer domain, as it is clearly ∅.

On the other hand, the interpretation of non-denoting terms as referring to "second class objects" is compatible with Husserl's description of imaginary numbers. This description emphasizes that imaginary objects "do not occur" in the model of the theory, because they "are not defined there, are not established by means of the axioms" [11, p. 433]. Consequently, strictly speaking, imaginary elements *do not exist*, so they could be fairly considered nonexistent objects.

However, the notion of "nonexistent object" is highly controversial. Some authors have argued that the concept itself is contradictory. From a nominalistic perspective, there is a conflict between assuming the "existence" of nonexistents objects and the well-known Occam's Razor. If one accepts the "existence" of these objects, then the existence of numbers, sets and every other mathematical entity, as well as fictional characters, should be also taken for granted. For this reason, the nonexistent values approach to non-denoting terms implies, I think, a weak form of platonism.

Farmer [8, p. 1274] also explained other difficulties of this approach. Firstly, he maintained that the conception of the referents of non-denoting terms as "second class objects" is contrary to mathematical practice, for the reason that, in ordinary mathematics, all the objects of a domain are under the scope of the quantifiers. And secondly, he remarked that bound and free variables do not denote exactly *the same kind of value*, what again seems a bit odd and unnatural with the glasses of ordinary mathematics.

Let me conclude this section with a summary of Centrone's objection to Da Silva [5]. While discussing this paper, she anticipated the inclusion of a unary predicate[8] in the object language to restrict quantification. She rightly argues that R is not a proper symbol of, for instance, the language of arithmetic:

> One might think of a monadic formula $R(x)$ of $\mathscr{L}$ defining the objects of **D**, so that $\mathscr{L}$**D**-formulae would in this case be just those formulae of $\mathscr{L}$ in which quantification is restricted to $R(x)$ [...] The problem is that such a $\mathscr{L}$-formula $R(x)$, in most cases, does not exist (for instance, if Γ is first-order Peano arithmetic PA and **D** is the standard model of PA, it is known that no $\mathscr{L}(\text{PA})$-formula $R(x)$ exists which defines the set $\mathbb{N}$ of standard natural numbers in every model of Γ).
> [3, p. 177]

Hence, Centrone criticized the attempt to "fix" the intended interpre-

---

[8]The unary predicate R is introduced in Da Silva [6], but it is not mentioned in Da Silva [5].

tation of a theory by means of a unary predicate R. Since $R(x)$ is not a well-formed formula of the language of arithmetic, Da Silva's solution seemed rather *ad hoc* to her. In fact, the necessity of introducing the predicate is not even mentioned in the *Doppelvortrag*.

3 The "Partial" interpretation

3.1 From partial valuation to free logics

Farmer defended that his preferred approach to deal with non-denoting terms is a partial valuation for terms, but total valuation for formulas. This position rejects the *existence assumption* of classical logic, that is, the belief that terms always have a denotation. Thus, if one adopts the "partial" perspective, non-denoting terms simply do not have a referent (its reference is not found neither in another *sort*, nor in an *outer domain* of nonexistent objects). "The mathematician normally thinks of a term such as $\frac{1}{0}$ as a legitimate expression which can be manipulated and reasoned with, but which absolutely has no denotation" [8, p. 1271].

Let $\mathfrak{J}$ be an interpretation such that $\mathfrak{J} = \langle \mathcal{A}, g \rangle$, and suppose that $\mathbf{A} = \mathbb{R}$. Since $i \notin \mathbb{R}$, the term $\sqrt{-1}$ must be non-denoting. Therefore, it is natural to believe that the interpretation function, whose domain for terms is TERM($\mathscr{L}$) (the set of terms of $\mathscr{L}$) has to be partial[9]. Expressed formally $\mathfrak{J} : \text{TERM}(\mathscr{L}) \longmapsto \mathbb{R}$. From this point of view, the transition from the reals to the complex numbers may be viewed as the process of extending the set of possible inputs for $\mathfrak{J}$, in parallel with the conversion of the square root function $\mathbf{f}$ into a total function. The new values of $\mathbf{f}$ are thus the referents of non-denoting terms when $Th(\mathcal{A})$ is enlarged. This partial valuation for terms is finally established by including the following semantical rule: a term has a denotation if all its subterms have a referent [8, p. 1274].

The possibility of extending the operations defined on a domain $\mathbf{A}$ is compatible with Husserl's characterization of "relative definiteness". As I have showed above, a relatively definite theory admits "for a broader domain the same, and then of course also new, axioms" [11, p. 426]. This is equivalent to say that, contrary to those which are absolutely

---

[9]$\mathfrak{J}(x) = g(x)$ and it is always defined, since variables must have a denotation.

definite, relatively definite theories do leave something open. "An axiom system can delimit a sphere of existence and leave open a vague, broader sphere" [11, p. 437]. By contrast, Husserl believed that it was not possible to extend, without contradiction, the operations described by an absolutely definite axiom system[10], as it "leaves for its operational substrate absolutely nothing open with respect to the operations defined" [11, p. 436]. Let me now quote the relevant passage by Husserl, where he is summarizing his own account of absolute definiteness:

> It leaves open no question whatsoever that the operation system offers, therefore also leaves no operational formations undefined and unregulated, and consequently admits of no expansion of the operational domain [...] No proposition meaningful in virtue of the axioms becomes false because of the fact that it has recourse to operational formations which are not defined. [11, p. 451]

From the last part of the quote it is evident that, in the language of an absolutely definite theory, there is no term $f(a)$ which is non-denoting for the reason that $\mathbf{f}$ is not defined for $\mathbf{a} \in \mathbf{A}$. On the contrary, in a relatively definite theory, $\mathfrak{J}(\sqrt{-1})$ must be undefined, due to the fact that the square root function is not defined for -1. In the –relatively definite-system of the "whole numbers, positive and negative", the equation $x^2 = -a$, $x = \pm\sqrt{-a}$ "certainly has a sense"[11] [11, pp. 438-39], but $\mathfrak{J}(\sqrt{-a}) \notin \mathbf{A}$. For this reason, Husserl concluded that this equation cannot hold "at all" in the considered system. Therefore, formulas that contain a non-denoting term must be false.

Once he established the partial valuation for terms, Farmer [8, p. 1274] imposed a total valuation for formulas. It follows that a formula φ always has a truth-value, even when φ includes non-denoting terms (the interpretation function for formulas, whose domain is FORM($\mathscr{L}$), is not partial). In the context of a *free logic*, which is a formal system that

---

[10]Notice that it is impossible to enlarge the *ordered field* of the reals without contradiction, for the reason that the complex numbers cannot be totally ordered.

[11]Farmer also discussed the position according to which every non-denoting term is not a well-formed expression of the language. However, this approach cannot be Husserl's, as for him $x^2 = -a$ "certainly has a sense" within an axiom system for the "whole numbers".

rejects the aforementioned "existence assumption", the semantics for formulas containing non-denoting terms can be positive (these formulas are true) or negative (they are false). Clearly, a *negative semantics* fits with Husserl's position in the *Doppelvortrag*. Farmer also adopts a "negative" semantics by introducing the second semantical rule: "a formula is false if any term occurring in it is non-denoting" [8, p. 1274].

I agree with Farmer when he claimed that this approach to non-denoting terms is the most natural[12]. For a mathematician, the formula $\sqrt{x} = 2$ is true for 4, but it is false for every x such that $x \neq 4$, what of course includes every y such that $y < 0$ (that is, all the values which do not have a real square root). "These statements are typical of the kind of assertions that mathematicians make using partial functions, and they are true on the basis of the two valuation rules given above" [8, p. 1275].

Now, consider the equation $1 + (-1) = 0$. It is very plausible to believe that the inference from $1 + (-1) = 0$ to $\exists x(1 + x = 0)$ should only be permitted in case that -1 had a denotation. With the tools of free logic at hand, it is possible to formalize this intuition. The logical symbols of the language of a free logic may include the usual connectives and quantifiers, together with equality and a unary *existence predicate*, $E!$ [13, p. 1024]. For every term t such that $t \in \text{TERM}(\mathscr{L})$, $E!(t)$ is true if t denotes an object under the scope of the existential quantifier; it is false otherwise. For this reason, $E!$ is definable as follows:

$$E!(t) := \exists x(x = t).$$

Therefore, the inference from $1 + (-1) = 0$ to $\exists x(1 + x = 0)$ can be weakened by imposing that $1 + (-1) = 0, E!(-1) \vdash \exists x(1 + x = 0)$. In a free logic, the "existence assumption" is no longer a presupposition, but a separated formula which could be satisfied by the referents of every denoting term in TERM($\mathscr{L}$). Consequently, the "transition through the Imaginary" may be regarded as the successive enlargement of a system by means of $E!$-sentences. This point of view is compatible with Peano's description[13] of the extension of the "concept of number":

---

[12]In fact, the partial valuation for terms and total valuation for formulas have often be considered the "Gold Standard" for dealing with non-denoting terms (see [FOM] free logic).

[13]As Detlefsen has pointed out, the following description by Gauss is historically

> There does not exist a number (from the sequence 0, 1, ...), which when added to 1 gives 0.
>
> There does not exist a number (integer), which multiplied by 2 gives 1.
>
> There does not exist a number (rational), whose square is 2.
>
> There does not exist a number (real), whose square is -1.
>
> Then one says: in order to overcome such an inconvenience, we extend the concept of number, that is, we introduce, manufacture, create (as Dedekind says) a new entity, a new number, a sign, a sign-complex, etc., which we denote by -1, or $\frac{1}{2}$, or $\sqrt{2}$, or $\sqrt{-1}$, which satisfies the condition imposed. [14, p. 224]

In the next section, I will show that, unlike the sort symbols and the predicates used to restrict quantification, the attribution of an "existence predicate" to Husserl is not *ad hoc*. There is textual evidence to conclude that what he called "existence axioms" play an important role in his solution to the problem of imaginary numbers.

3.2 The "existence axioms" and Husserl's solution

Husserl's existence axioms are not mentioned neither in Da Silva [5] nor in Da Silva [6]. However, these axioms are included in every theory that has an intended model. "Any axiom system must –in order that one can say of it at all [..] that it defines a manifold– include existence axioms" [11, p. 420]. Husserl defended that, if the sum can be totally defined on a domain, then, for two particular numbers **a** and **b** of **A**, there must be an object $\mathbf{x} \in \mathbf{A}$ such that **x** is the result of $\mathbf{a} + \mathbf{b}$. Hence, in the language of a free logic, $E!(a+b) := \exists x(x = a+b)$. Furthermore, the axioms of this form are essential in the construction of the successive "levels" of numbers:

---

more accurate than Peano's: "Starting originally from the notion of absolute integers, it has gradually enlarged its domain. To integers have been added fractions, to rational quantities the irrational, to positive the negative, and to the real the imaginary" [7, p. 279].

> The distinction between the levels resides in the existence axioms; the existence of particular forms of operation is stipulated under narrower or broader conditions. [11, p. 448]
>
> The distinction between axiom systems lies in the existence axioms which are either broader or narrower. [11, p. 449]

It is thus clear that the "broadest" set of existence axioms is the one associated to the theory of the complex numbers, while the "narrowest" one is the collection of existence axioms for the natural numbers. The reason is that the equations $1+x = 0$, $2 \cdot x = 1$, $x^2 = 2$ and $x^2 = -1$ have a solution in the field of complex numbers. As a result, the following $E!$-formulas are all satisfied by a mathematical structure whose domain is $\mathbb{C}$:

$$E!(-1) := \exists x(x = -1),$$
$$E!(\tfrac{1}{2}) := \exists x(x = \tfrac{1}{2}),$$
$$E!(\sqrt{2}) := \exists x(x = \sqrt{2}),$$
$$E!(\sqrt{-1}) := \exists x(x = \sqrt{-1}).$$

The process of enlarging a system by means of $E!$-sentences is, equivalently, the process of extending its class of existence axioms. In my view, these sentences are a very natural way to express formally Husserl's notion of "existence axiom". In fact, the interpretation of the transition through the Imaginary as a progressive increase[14] of existence axioms is supported by the passages quoted above. Thus, starting with the theory of natural numbers, the extension of the "concept of number" *maximise* the number of existence axioms that are true in the enlarged systems (or, similarly, the number of equations that have a solution).

The concept of "existence axiom" also allows to revisit the meanings of relative and absolute definiteness. A relatively definite theory can be understood, from this point of view, as a theory that *admits* new existence axioms; an absolutely definite one, as a theory to which "no further axiom which could be added" [11, p. 427]. In addition to this, it should be remarked that, if the axioms of $Th(\mathcal{A})$ are restricted to the set of existence axioms, then it is true that "for a broader domain the

---

[14]"Expansion by means of existence axioms, and thus expansion of the domain (within the sphere of the same operations)" [11, p. 477].

same, and then of course also new, axioms are valid" [11, p. 426]. It is a simple truth of elementary model theory that existentially quantified formulas are always preserved under extension[15] and expansion[16] of a mathematical structure $\mathcal{A}$.

Consider the sentence $\exists x_1, ..., x_n \Psi(x_1, ..., x_n)$ and assume that it is true in $\mathcal{A}$. It follows that there must be (at least) a n-tuple $\langle \mathbf{x_1}, ..., \mathbf{x_n} \rangle$ in $\mathbf{A} \times \overset{n}{\cdots} \times \mathbf{A}$ such that $\langle \mathbf{x_1}, ..., \mathbf{x_n} \rangle$ satisfies Ψ. Since the extension $\mathcal{B}$ of $\mathcal{A}$ implies that $\mathbf{A} \subseteq \mathbf{B}$, it is evident that the same n-tuple $\langle \mathbf{x_1}, ..., \mathbf{x_n} \rangle$ is also in $\mathbf{B} \times \overset{n}{\cdots} \times \mathbf{B}$. Thus, $\exists x_1, ..., x_n \Psi(x_1, ..., x_n)$ is true in $\mathcal{B}$. (The proof is continued analogously for the case of the expansion $\mathcal{B}$ of $\mathcal{A}$.)

The fact that the n-tuple $\langle \mathbf{x_1}, ..., \mathbf{x_n} \rangle$ satisfying

$$\exists x_1, ..., x_n \Psi(x_1, ..., x_n)$$

is included in *all* the extensions and expansions of $\mathcal{A}$ explains why Husserl believed that "every number domain of lower level is completely contained[17] in every number domain of higher level" [11, p. 448]. He also defended that all the operations defined on a "level" are also defined in higher levels of the hierarchy of numbers. Similarly, Husserl thought that "every axiom system of lower level is completely contained in every axiom system of higher level" [11, p. 448]. This can be justified only if the considered "axiom system" is a set of existence axioms (clearly, the universally quantified sentences are not necessarily preserved under extension and expansion of $\mathcal{A}$). Surprisingly, Husserl claimed that $Th(\mathcal{A}) \subseteq Th(\mathcal{B})$ is a necessary condition for the enlargement of a system of numbers:

---

[15]"If $\mathcal{A}$ and $\mathcal{B}$ are L-structures with $\mathbf{A} \subseteq \mathbf{B}$ and the inclusion map $i : \mathbf{A} \longrightarrow \mathbf{B}$ is an embedding, then we say that $\mathcal{B}$ is an extension of $\mathcal{A}$" [10, p. 6].

[16]"Suppose L^- and L^+ are signatures, and L^- is a subset of L^+. Then if $\mathcal{A}$ is an L^+-structure, we can turn $\mathcal{A}$ into an L^--structure by simply forgetting the symbols of L^+ which are not in L^- [...] The resulting L^--structure is called the L^--reduct of $\mathcal{A}$ or the reduct of $\mathcal{A}$ to L^- [...]

When $\mathcal{A}$ is an L^+-structure and $\mathcal{C}$ is its L^--reduct, we say that $\mathcal{C}$ is an expansion of $\mathcal{C}$ to L^+" ([10, p. 9].

[17]I have recently argued [1] that, according to Husserl, the highest levels of the hierarchy of numbers must contain *a copy* of the previous levels, what is essential to understand his solution to the problem of imaginary numbers. For my own view on Husserl's notions of completeness, see also Aranda [2].

> Precisely the relationship between axiom systems, according to which a narrower is contained within a broader, an essential one within a still more essential one, is the presupposition for the possibility of the transition through the Imaginary. [11, p. 451]

In my view, this passage should be read with the notion of "existence axiom" in mind. Let $\mathcal{B}$ be an extension (or an expansion) of $\mathcal{A}$, and $E!(t)$ an existence axiom of $Th(\mathcal{A})$. Since all the existentially quantified sentences are preserved from $\mathcal{A}$ to $\mathcal{B}$, it is evident that $E!(t) \in Th(\mathcal{A})$ implies $E!(t) \in Th(\mathcal{B})$. Now, if the existence axiom $E!(t)$ makes possible the inferences about the referent of t (just like the $E!$-formulas of a free logic), then the inferences about $\mathfrak{J}(t)$ can be carried out in both $Th(\mathcal{A})$ and $Th(\mathcal{B})$. In other words, the sentence $\exists x(x \cdot x = 2)$ follows logically from the theory of complex numbers for the reason that the existence axiom $E!(\sqrt{2}) := \exists x(x = \sqrt{2})$ is included in this theory (notice that $E!(\sqrt{2})$ is included in every axiom system of "higher level", but not in "lower levels").

Thus, the logical consequences of the existence axioms are also preserved under extension and expansion. The sentence $\exists x(x \cdot x = 2)$ is "permitted" in $Th(\mathcal{B})$ (that is, in the theory of complex numbers) because its truth depends only on the existence axioms of $Th(\mathcal{A})$ (of the theory of real numbers):

> The inference from the imaginary is permitted in the singular case or for a class, if we can know in advance and can see that for this case or for this class the inference is decided by the narrower system. [11, p. 437]

And, if the operations between "old" numbers in the enlarged domain depend on the laws for the starting system, then the addition of "new" numbers (the "transition through the Imaginary") can never lead to contradictions.

4 Conclusions

This paper focuses on modern interpretations of Husserl's solution to the problem of imaginary numbers, which he addressed in his 1901 *Dop-*

pelvortrag. The extension of the concept of number leads to new entities (-1, $\frac{1}{2}$, $\sqrt{2}$, $\sqrt{-1}$, etc.) that are the referents of (previously) non-denoting terms. The question is how to guarantee that this extension does not imply a contradiction.

One possible answer is to argue that the quantifiers of a sentence like, for instance, $\exists x(x = \sqrt{-1})$ range over different universes ("sorts") or over different kind of values. In the first case, the formal language must contain at least two sort symbols or two unary predicates, depending on whether the structure is many-sorted or one-sorted. It is evident that $\neg\exists x(R_\sigma(x) \wedge x = \sqrt{-1})$ does not contradict $\exists x(R_\rho(x) \wedge x = \sqrt{-1})$. An important defect of this approach is the proliferation of sort symbols (or unary predicates). In the second case, the language must contain a unique unary predicate, what has been defended by Da Silva and criticized by Centrone. The predicate allows to restrict quantification to a subset **D** (the "inner" domain of existent objects) of the domain **A** (the union of the inner domain with an "outer" domain of *nonexistent* values). Clearly, the truth of $\neg\exists x(R(x) \wedge x = \sqrt{-1})$ in $\mathcal{A}$ (whose inner domain is $\mathbb{R}$) does not contradict the truth of $\exists x(R(x) \wedge x = \sqrt{-1})$ in $\mathcal{B}$ (whose inner domain is $\mathbb{C}$), because $R^\mathcal{A} \neq R^\mathcal{B}$. A serious difficulty of this perspective is that it presupposes the existence of "second class" objects, which are a bit odd and unnatural from the point of view of ordinary mathematics.

Another possible answer is to maintain that $\exists x(x = \sqrt{-1})$ should be considered as "existence axiom". In the context of free logic, sentences of this form (which are defined as $E!(\sqrt{-1})$) allow the inferences about the complex number i. An existence axiom states explicitly that a term t has a denotation, since the *existence assumption* of classical logic is rejected in general. Furthermore, a partial interpretation for terms and a total valuation for formulas is compatible with Husserl's claims in the Double Lecture, and it is also closer to mathematical practice. Although Husserl's existence axioms have not been fairly emphasized in the literature, they played an important role in his solution. The textual evidence suggests that Husserl thought that *all* the (existence) axioms for a "low level" of the hierarchy of numbers should be included in higher levels. He believed that this fact was a necessary condition for the "transition through the Imaginary". Since every existentially quantified sentence

$E!(t)$ must be preserved under the extension (or expansion) $\mathcal{B}$ of $\mathcal{A}$, it follows that $E!(t) \in Th(\mathcal{B})$. Obviously, if $E!(t) \vdash \varphi$, then it also holds that $\varphi \in Th(\mathcal{B})$, what, for Husserl, solves the problem of imaginary numbers.

References

[1] Aranda, V. Completeness, Categoricity and Imaginary Numbers: The Debate on Husserl, *Bulletin of the Section of Logic*, 49(2), 109-125, 2020.

[2] Aranda, V. Completeness: From Husserl to Carnap, *Logica Universalis*, https://doi.org/10.1007/s11787-021-00283-4, 2021.

[3] Centrone, S. *Logic and Philosophy of Mathematics in the Early Husserl*. Springer, Berlin, 2010.

[4] Chang, C. and Keisler, H. J. *Model Theory*. Dover publications, New York, 2012.

[5] Da Silva, J. J. Husserl's two notions of completeness. *Synthese*, 125(3), 417–438, 2000.

[6] Da Silva, J. J. Husserl and Hilbert on completeness, still. *Synthese*, 193(6), 1925–1947, 2016.

[7] Detlefsen, M. Formalism. In: Saphiro, S. (ed.) *The Oxford Handbook of Philosophy of Mathematics and Logic* (pp. 236–317). Oxford University Press, Oxford, 2005.

[8] Farmer, W. M. A partial functions version of Church's simple theory of types. *The Journal of Symbolic Logic*, 55(3), 1269–1291, 1990.

[9] Hartimo, M. Husserl on completeness, definitely. *Synthese*, 195(4), 1509–1527, 2018.

[10] Hodges, W. *Model Theory*. Cambridge University Press, Cambridge, 1993.

[11] Husserl, E. *Philosophy of arithmetic: Psychological and logical investigations with supplementary texts from 1887-1901*. Springer Science+Business Media, Dordrecht, 2003.

[12] Manzano, M. *Extensions of first-order logic*. Cambridge University Press, Cambridge, 1996.

[13] Nolt, J. Free logics. In: Gabbay, D. M. et al (eds.) *Philosophy of logic*, pp. 1023-1060. North-Holland, Amsterdam, 2007.

[14] Peano, G. Foundations of analysis. In: Kennedy, H. (ed.) *Selected Works of Giuseppe Peano* (pp. 219–226). University of Toronto Press, Toronto, 1973.

THE MODEL-THEORETIC SQUARE OF OPPOSITION

RYAN CHRISTENSEN
Brigham Young University
ryan.christensen@byu.edu

Aristotle introduces the square of opposition in *De Interpretatione* 17b16–26 by defining the relevant terms—traditionally translated 'contradictory' and 'contrary'—syntactically, in terms of the form of the sentence. He then goes on to indicate some semantic relations that he takes to follow from these definitions: exactly one of a pair of contradictories is true; contraries cannot both be true; subcontraries can both be true.[1] This, for example, is what he says about contraries:

> I call the universal affirmation and the universal negation *contrary* opposites, e.g. 'everyone is just'—'no one is just'. So these cannot be true together, but their [contradictory] opposites may both be true.[2]

If we consider more broadly what Aristotle says about contraries, it seems clear that he had a stronger semantic relation in mind than mere incompatibility. His examples of contraries are always stark contrasts: all and none, black and white (see, for example, *Categories* 4a10–22). Mere incompatibility is had by each of the following pairs:

This paper was presented at the workshop in Rio; thanks to the participants of that workshop. Thanks also to the anonymous referee for very useful feedback.

[1] Aristotle does not use the term 'subcontrary', but he does claim that the contradictories of contraries can both be true, as can be seen in the following quotation.

[2] All quotations from Aristotle are from [1]. This quotation has been altered slightly.

All tigers are fierce
Necessarily, p
John and Jane are both tall
There are exactly two books on the table

Tabitha the tiger is not fierce
Not-p
John is not tall
There are exactly four books on the table

These are pairs of contraries in the traditional sense, yet the examples that Aristotle gives us have a stronger property than contrariness, a property we might call *antithesis*. The antithesis of 'All tigers are fierce' is not simply any sentence that is incompatible with it, but its polar opposite: 'No tigers are fierce'. Similarly, the antithesis of 'Necessary, ϕ' is 'Necessarily, not-ϕ' and the antithesis of 'John and Jane are both tall' is 'Neither John nor Jane is tall'. It seems plausible that not every sentence has an antithesis—there seems to be no antithesis to 'There are exactly two books on the table'—but if a sentence has an antithesis, it is unique.

This seems to be Aristotle's considered doctrine at the time of *De Interpretatione*: that not every property and proposition has a contrary, but that if a property or proposition has a contrary, it has only one. He repeatedly uses superlative terms to describe contraries, such as this: "We call contraries ... the most different of the things in the same genus" (*Metaphysics* Δ.10, 1018a25–28. See also a similar passage in *DI* 23b23–24.)[3]

But this claim of uniqueness troubled him; it is among the list of metaphysical aporias that he poses in the first book of the *Metaphysics*. He ultimately concludes that "one thing cannot have more than one contrary" (*Met.* 1055a19), since contraries express superlative difference. This uniqueness doctrine is difficult to reconcile with the theory in the *Nicomachean Ethics* that virtue is a mean, since this theory requires that a property (e.g., temperance) have more than one contrary (abstemiousness and gluttony). But even here he wants to distinguish the true contrary from what we might call the quasi-contrary:

> These states being opposed to one another, the greatest contrariety is that of the extremes to each other, rather than

---
[3]This passage in the *Metaphysics* has five different definitions of 'contrary'. The first is closest to the traditional semantic definition, but it also includes a metaphysical element: the properties must be "in the same genus."

The Model-Theoretic Square of Opposition

to the intermediate.... [N]ow contraries are defined as the things that are furthest from each other, so that things that are further apart are more contrary. (*NE* II.8, 1108b26–35)

Because the vices are most opposite, they are true contraries; the virtue is contrary to the vices only in the weaker sense of incompatibility.[4]

I think the best approach is to say that 'contrary' might be understood semantically, in which case a proposition will, in general, have many contraries. Or it might be understood in a stricter sense, in which it means 'antithesis'—a unique, superlative opposite. Model theory helps bring out this stricter sense.

This paper explores the logic of antitheses. It will focus on antitheses for the modern square of opposition, which describes the relation between *inner negation* and *outer negation*, defined syntactically. Given any pair of appropriate dual operators, we can form a square; for example:

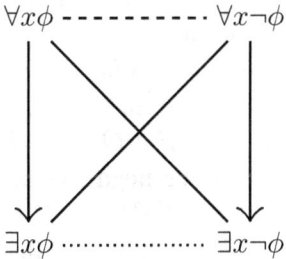

Since $\forall x \neg \phi \equiv \neg \exists x \phi$, the square maps out the relations between inner and outer negation.[5]

---

[4] [3] takes the two vices to be true contraries, occupying the top two corners of a square of opposition, and the corresponding virtue to be the conjunction of the bottom two corners. This approach takes the vices to be given by the rules "Always do x" and "Never do x," and the virtues to be given by the moderate rule "Sometimes do x and sometimes don't." This solution makes the virtue a semantic contrary without being an antithesis.

[5] See [5]. A *traditional* square takes these relationship to hold among the categorical statements, e.g., "Every A is B" and "Some A is B." Given the standard translation of the sentences as $\forall x(Ax \to Bx)$ and $\exists x(Ax \land Bx)$, subalternation does not hold. There are other operators which likewise fail subalternation: One example is $\Box$ and $\Diamond$ on weak logics, such as K. As another example, [4] cites *already* and

253

Thus I am drawing in this paper the connections between (on one hand) the *syntactic* notion of inner and outer negation, and (on the other hand) various *model-theoretic* notions. I will focus on three model-theoretic notions: *complement*, *antithesis*, and *dual*, which will be defined and illustrated below. These match up with the syntactic notions of outer negation, inner negation, and dual.

The main question of this paper is this: Is there a perfect model-theoretic analogue of the syntactic notions? Inner and outer negations can be defined uniquely, such that any formula of a given logic has exactly one inner and one outer negation. Is it possible to define a model-theoretic analogue? If so, we have a model-theoretic account of antithesis.

1 Truth Functions

Let's begin with truth functions.

Syntactically, a TF sentence is defined by the usual recursive rules. The *outer negation* of a sentence ϕ is $\neg\phi$. So, for instance, the outer negation of $(P \wedge Q)$ is $\neg(P \wedge Q)$. The *inner negation* of ϕ is generated by substituting, for each atomic sentence α in ϕ, the sentence $\neg\alpha$. So, for instance, the inner negation of $(P \wedge Q)$ is $(\neg P \wedge \neg Q)$. The (syntactic) *dual* of a sentence ϕ is the outer negation of the inner negation of ϕ. Hence the dual of $(P \wedge Q)$ is $\neg(\neg P \wedge \neg Q)$. Because this sentence is semantically equivalent to $(P \vee Q)$, $\vee$ is called the dual operator of $\wedge$. It is easy to see that, modulo double negation, duality is symmetric: if ϕ is the dual of ψ, then ψ is the dual of ϕ.

Model-theoretically, a sentence can be represented by a bitstring, so that for instance the sequence 1000 represents $(P \wedge Q)$.[6] If the language is given as an ordered set of atomic sentences, the bitstring unambiguously represents a certain sentence. Bitstring notation can be seen as a compact way of expressing the standard truth tables for the truth-functional operators.

---

still, distinguishing between an *Aristotelian* square, defined by the presence of these semantic relationships, and a *duality* square, defined syntactically. This paper focuses on duality squares, but only in cases in which the semantic relations also hold.

[6]For another use of bitstrings in approaching Aristotelian relations, see [6].

The *complement* of a sentence is the one's complement of the bit-string, which replaces every 0 with 1 and every 1 with 0. The *reverse* of a sentence is the sentence that results from reversing the order of the bits (e.g., the reverse of 1000 is 0001). The (model-theoretic) *dual* of a sentence is the complement of its reverse. Thus the dual of 1000 is 1110. It is, once again, easy to see that duality is symmetric.

One virtue of this notation is that it makes clear the connection between the syntax and the model theory. The outer negation of a sentence is its complement, since the truth-table semantics of negation is one's complement. The inner negation of a sentence is its reverse, since inner negation has the semantic effect of reversing the polarity of the arguments of truth functions, which has the effect of reversing the values. The dual, being the outer negation of the inner negation, is thus the complement of the reverse.

For example:

sentence	$(P \wedge Q)$	1000
outer negation	$\neg(P \wedge Q)$	0111
inner negation	$(\neg P \wedge \neg Q)$	0001
dual	$\neg(\neg P \wedge \neg Q)$	1110

The main question of this paper is whether there are model-theoretic analogues of the syntactic relations. It is clear that there is one for duality, but antithesis does not show up on this list. The reverse of a sentence is not the same as its antithesis, and clearly not the same as its contrary. I take the traditional definition of contrary to have two requirements: ψ is a (semantic) *contrary* of a sentence ϕ iff (1) in every bit where ϕ has 1, ψ has 0; and (2) there is at least one bit where ϕ and ψ both have 0.[7] The second clause is needed to distinguish the contrary from the contradictory. So, for instance, 1000 has seven contraries: 0110, 0101, 0100, 0011, 0010, 0001, or 0000. There will typically be many contraries, as there are in this case. In other cases (tautologies) there are none.[8] But of course the reverse of a given sentence is unique.

---

[7] This is a specification for bitstrings of the definition "ϕ and ψ are contraries if they can't both be true but can both be false."

[8] The number of contraries will be $2^n - 1$, where n is the number of instances of 0 in the bitstring.

255

However, in special cases, the reverse is the best candidate for the antithesis. An antithesis is supposed to be a certain kind of contrary; it is also supposed to be the kind of polar opposite that the reverse exemplifies. So we take the antithesis to be both: ψ is the *antithesis* of ϕ iff (1) ψ is the reverse of ϕ, and (2) ψ is a contrary of ϕ. In the example above, the sentence $(P \wedge Q)$, which is represented by the bitstring 1000, has as its reverse 0001, which is also a contrary. Hence, by this definition, 0001 is the antithesis of 1000.

We can also define *subantithesis*, which stands to subcontrary as antithesis stands to contrary. We say that ψ is a (semantic) *subcontrary* of a sentence ϕ iff (1) in every bit where ϕ has 0, ψ has 1; and (2) there is at least one bit where ϕ and ψ both have 1. Then ψ is the *subantithesis* of ϕ iff (1) ψ is the reverse of ϕ, and (2) ψ is a subcontrary of ϕ.

Some sentences have antitheses; others have subantitheses. The reverse of 1000 is 0001; these sentences are antitheses. The reverse of 1110 is 0111; these sentences are subantitheses. Some sentences have neither: the reverse of 1100 is 0011, which is neither an antithesis nor a subantithesis. If a sentence has an antithesis (subantithesis), it has a unique (sub)antithesis, because the reverse of a sentence is unique. And since both reverses and contraries are symmetric, (sub)antitheses are symmetric.

Given this, we can produce an entire square given any corner, modulo the left-right symmetry. For example, consider the sentence 1000. This sentence has an antithesis (0001), so these two sentences must take the A and E corners. Their complements can be computed (0111 and 1110, respectively), and the entire square constructed. Then consider 1110. This has no antithesis, but it has a subantithesis. So 1110 and its subantithesis 0111 must be the I corner and the O corner. Their complements (0001 and 1000) are the diagonal A and E. If a sentence has neither an antithesis nor a subantithesis, its reverse will also be its complement, and no square is possible. In such a case, the A and I sentences will be the same, as will the E and O. For example, let A be 1100. Its dual (the complement of its reverse) is also 1100. Thus $A = I = 1100$, and $E = O = 0011$.

I end this section with one more example of a square, this time on a language with three atomic sentences $\langle P, Q, R \rangle$:

The Model-Theoretic Square of Opposition

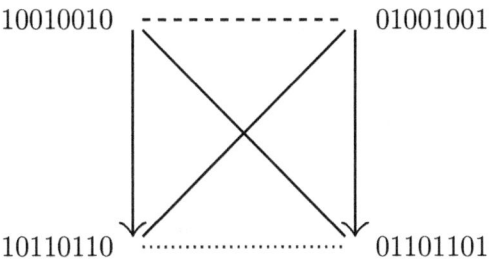

The A sentence can be determined by the bitstring to be $(P \wedge Q \wedge R) \vee (P \wedge \neg Q \wedge \neg R) \vee (\neg P \wedge \neg Q \wedge R)$. The other corners can likewise be calculated and verified.

2 Monadic First-Order Logic

I turn now to *monadic first-order logic* (FOL_1), which is a restriction of first-order logic.

We start, as before, with the syntax. A language $\mathcal{L}$ is an ordered set of monadic predicates; it contains no constants. Given a language, a *well-formed formula* (wff) can be defined in something like the usual way. The important restriction here over the standard rules is that a wff must have only a single quantifier, and it must be the first symbol in the wff. A sentence that begins with a universal quantifier will be called a *universal sentence*; a sentence that begins with an existential quantifier, an *existential sentence*.

Given the restriction that every sentence must begin with a quantifier, $\neg \forall x.\phi$ is not a wff; it is, however, equivalent to $\exists x.\neg \phi$.[9] And so we can define *outer negation* like this:

> The outer negation of $\forall x.\phi$ is $\exists x.\neg \phi$
> The outer negation of $\exists x.\phi$ is $\forall x.\neg \phi$

Inner negation can likewise be defined:

> The inner negation of $\forall x.\phi$ is $\forall x.\neg \phi$
> The inner negation of $\exists x.\phi$ is $\exists x.\neg \phi$

---

[9] To avoid a distracting proliferation of brackets, I use a dot to indicate the scope of the quantifier.

257

(Syntactic) *duality* can be defined:

> The dual of $\forall x.\phi$ is $\exists x.\phi$
> The dual of $\exists x.\phi$ is $\forall x.\phi$

Turning to the model theory, given a language $\mathcal{L}$ of length s, a *model* consists of a non-empty *domain* D of up to 2^s objects, and, for each $P \in \mathcal{L}$, a subset of D. Models are individuated by their members, so that, for a language of size s, there are $2^{2^s} - 1$ models. For example, we can list the models of a language $\mathcal{L} = \langle P, Q \rangle$ like this:

P	Q	15	14	13	12	11	10	9	8	7	6	5	4	3	2	1
1	1	1	1	1	1	1	1	1	1	0	0	0	0	0	0	0
1	0	1	1	1	1	0	0	0	0	1	1	1	1	0	0	0
0	1	1	1	0	0	1	1	0	0	1	1	0	0	1	1	0
0	0	1	0	1	0	1	0	1	0	1	0	1	0	1	0	1

The top row of the table indicates whether the model has an object that is in the subsets of both P and Q, the second, of P but not Q, and so on. I will indicate models by the bitstring associated with the model; for instance, the model with only a single object in the subsets of P and Q (model 8 on the table), is indicated by 1000. This terminology makes it easy to determine whether one model is a submodel of another: $\mathcal{A} \subseteq \mathcal{B}$ iff every bit where $\mathcal{A}$ has a 1, $\mathcal{B}$ also has a 1.

A *query* $\mathcal{Q}$ is a set of models. A query *defined* by a sentence ϕ, denoted $\mathcal{Q}(\phi)$, will be the set of models of ϕ. Let $\top$ be the universal query, the set of all models of $\mathcal{L}$, and $\bot$ the empty query.

The *complement* of a query $\mathcal{Q}$ is $\top - \mathcal{Q}$, the set of all models not in $\mathcal{Q}$. It is easy to see that complementation is the model-theoretic analogue of outer negation: $\mathcal{Q}(\neg \phi) = \top - \mathcal{Q}(\phi)$.

Antithesis is intended to be the model-theoretic analogue of inner negation for universal sentences, so that the antithesis of $\mathcal{Q}(\forall x.\phi)$ is $\mathcal{Q}(\forall x.\neg\phi)$. Likewise, the subantithesis of $\mathcal{Q}(\exists x.\phi)$ is $\mathcal{Q}(\exists x.\neg\phi)$. Similarly, (model-theoretic) duality is intended to be the analogue of (syntactic) duality, so that the dual of $\mathcal{Q}(\forall x.\phi)$ is $\mathcal{Q}(\exists x.\phi)$.

The main question of the paper, once again, is this: Is there a model-theoretic relation between the antitheses and duals? The model-theoretic relation between complements is simple, so that it is a simple matter to go from any query to the query of its complement without calculating the sentence that defines the query. Is there such a method for antitheses and duals in FOL_1?

In unrestricted FOL, there is no such relation between queries. If we take the inner negation of a sentence $\forall x_1 \ldots x_n.\phi$ to be $\forall x_1 \ldots x_n.\neg\phi$, there are non-equivalent sentences that share an inner negation. For example, the empty query $\bot$ can be defined by any contradiction. So $\mathcal{Q}(\forall xy.Rxy \land \neg Ryx) = \mathcal{Q}(\forall xy.Rxy \land \neg Rxy)$, since $Raa \land \neg Raa$ is an instance of both sentences. But their inner negations are $\forall xy.Rxy \to Ryx$, the symmetric query, and $\forall xy.Rxy \to Rxy$, the universal query. Since a single query—the empty query—can be defined by more than one sentence, each of which has non-equivalent inner negations, one cannot get from a query alone to its antithesis.

The answer to the question, to be described in the next section, is Yes. In FOL_1, there are such relationships. Given a suitable query, it is possible to find its antithesis and dual.

3 Monadic Queries

We know that every monadic language has a maximum structure size: If the language is of size s, the largest model is size 2^s. For example, if the language is $\langle P, Q \rangle$, there will be a model with four objects: one with both P and Q, one with P but not Q, one with Q but not P, and one with neither.[10] Call the largest model for a language the *maximal model* for that language.

Object a in model $\mathcal{A}$ and object b in model $\mathcal{B}$ *correspond* iff a and b share all properties. Models $\mathcal{A}$ and $\mathcal{B}$ of a given language are *inverses* iff for each object in the maximal model, there is an object in exactly one of $\mathcal{A}$ and $\mathcal{B}$ that corresponds with it.

The bitstring notation makes it easy to find inverses: The inverse of a query is its one's complement. Thus FOL_1 inverses are structurally similar to TF complements. Another way to find the inverse of a model, given the standard ordering of models, is to subtract the model number

---

[10]That is, this is the largest model that can be distinguished by FOL_1. Models might be larger by duplicating one of the objects, but such a model cannot be distinguished by any sentence in the language.

from $2^s - 1$ (for languages of length s). Thus, for $s = 4$, the inverse of model 2 is model 13, and vice versa.

The next notion we'll need is a *plenary model*: The *plenary model* of a query Q is the model A such that for every B, $B \in Q$ iff $B \subseteq A$. A plenary model is defined as the unique such model. Not every query has a plenary model. For example, $\exists x.Px \lor Qx$ will not have a plenary model.

If we consider the query defined by $\forall x.Px \lor Qx$—$\{2, 4, 6, 8, 10, 12, 14\}$ — we see that one of these models, 14, has three objects; all the others have one or two. Each of the objects in the smaller models corresponds to one of the objects in model 14. Thus the plenary model of this sentence is simply 14. This single model encodes within it every way to make the sentence true.

Another way to visualize plenary models is with the following graph of all models of some language with two monadic predicates:

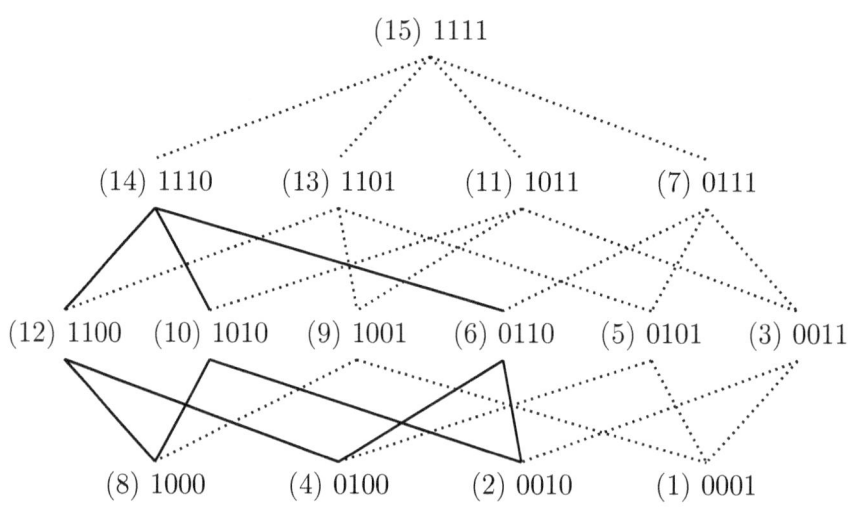

Each model is represented by a bitstring as in the above truth table, and numbered as in that table. The lines connecting models are submodel relations. For example, model 14 has three submodels of size 2 (6, 10, 12), each of which has two submodels of size 1. These relations are indicated by the solid lines in the diagram. Thus this is a diagram of

$\mathcal{Q}(\forall x.Px \lor Qx)$. The plenary model is the apex of the subgraph indicated by the solid lines.

It is also possible to represent universal sentences in FOL_1 as bitstrings. We first need a canonical way to state the sentence—a normal form. I will use *full disjunctive normal form* (full DNF), which lists the clauses of literals in standard order.

A *literal* is either an atomic sentence or a negation of an atomic sentence. A *clause* is a conjunction of literals. A formula ϕ in a language of size s is in *full DNF* iff

1. ϕ begins with a universal quantifier.

2. ϕ is a disjunction of one or more clauses of literals P_1 through P_s in order.

3. The clauses of ϕ are in standard order, which is the conventional order of rows on a truth table.

For a language $\mathcal{L}$ of size s, the largest full DNF will have 2^s clauses. Each of these clauses could be numbered, and each shorter full DNF can be uniquely specified by indicating which clauses it contains. Similarly, it could be indicated by a bitstring with a 1 indicating an included clause and a 0 indicating an excluded clause.

For example, in the language $\langle P, Q \rangle$, the largest full DNF is

$$\forall x.(Px \land Qx) \lor (Px \land \neg Qx) \lor (\neg Px \land Qx) \lor (\neg Px \land \neg Qx)$$

The formula $\forall x.Px \lor Qx$ in full DNF is

$$\forall x.(Px \land Qx) \lor (Px \land \neg Qx) \lor (\neg Px \land Qx)$$

This formula contains the first three clauses of the largest full DNF but not the fourth, so it could be expressed as 1110.

Using bitstrings to designate both sentences and models demonstrates the close connection between the syntax and the model theory. But the connection is not direct. This method allows only universal sentences to be designated by bitstrings; it says nothing about existential sentences. And the closest model-theoretic entity to a sentence is not a model but a query, as queries are uniquely determined by sentences. The connection is, rather, between a sentence and its plenary model: a single bitstring designates both a universal sentence and the plenary model of the query defined by that sentence.

In fact, every query defined by a universal sentence has a plenary model. Any universal sentence ϕ can be described by a bitstring that encodes the different ways of satisfying the sentence. This bitstring also designates a model $\mathcal{A}$. Every submodel of $\mathcal{A}$ also satisfies ϕ, since it satisfies at least one of the disjunctive clauses that are encoded in the bitstring. Every model that is not a submodel of $\mathcal{A}$ contains at least one object that is a counterexample to the universal.

This result is interesting because it allows a purely model-theoretic method for distinguishing universal queries from all other queries. It is easy to come up with examples of existential queries that do not have plenary models—for example, $\exists x.Px$ has 1111 as a model, but not 0011. But existential queries are not unique in this regard. A query is any set of models, and there are many queries—$\{8, 12, 14\}$, for example, or $\{4, 5, 13\}$, or $\{1, 2, 4, 8\}$—that do not have plenary models. In fact, there are only $2^{2^s} - 1$ models with plenary models for a language of size s (since there are that many models to serve as plenary models), but $2^{2^{2^s}-1}$ queries (since the set of queries is the power set of the universal query). Hence nearly all of the queries are not universal. Since every existential query is the complement of a universal query, there are the same number of existential queries as universal. Thus most queries cannot be defined by a formula in FOL_1.

The *expansion* of $\mathcal{A}$ is the set of all submodels of $\mathcal{A}$. It is easy to see that if $\mathcal{Q}$ is the expansion of $\mathcal{A}$, then $\mathcal{A}$ is the plenary model of $\mathcal{Q}$. $\mathcal{Q}_1$ is the *full inverse* of $\mathcal{Q}_2$ iff $\mathcal{Q}_1$ is the expansion of the inverse of the plenary model of $\mathcal{Q}_2$. It is easy to see that the full inverse relation is symmetric: If $\mathcal{Q}_1$ is the full inverse of $\mathcal{Q}_2$, then $\mathcal{Q}_2$ is the full inverse of $\mathcal{Q}_1$.

With these definitions, we can answer the question: Antitheses are full inverses. That is, $\mathcal{Q}(\forall x.\neg\phi)$ is the full inverse of $\mathcal{Q}(\forall x.\phi)$. And duals are complements of full inverses.

This method can be seen as reducing the problem of FOL_1 antitheses and duals to its TF counterpart. A plenary model for a universal

The Model-Theoretic Square of Opposition

sentence has the same bitstring as the corresponding TF sentence—e.g., the plenary model of $\forall x.Px \vee Qx$ has the bitstring 1110, which is the same as $P \vee Q$. And the inner negation of a FOL_1 sentence matches up with the outer negation of a TF sentence, since what is "inner" with respect to the quantifier is "outer" with respect to the rest of the sentence. And because FOL_1 inverses are structurally similar to TF complements, taking the inverse of the plenary model is like taking the complement in TF. Hence, this method, in a sense, bypasses the quantifier to make a TF sentence, finds the outer negation of that, and then adds the quantifier back.

This result shows that there is a model-theoretic analogue of inner negation for universal sentences. Because there is also a model-theoretic analogue of outer negation, we can construct a square of opposition (modulo left-right symmetry) given any query. For example, given the query $\{1, 2, 3, 4, 5, 6, 7\}$, which has 7 as its plenary model, we can find its antithesis, its complement, and its dual. It can easily be seen that this gives us the results we expect.

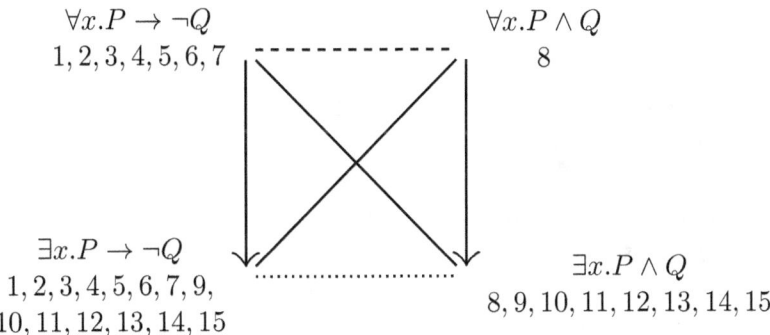

In this picture, the sentences that define the various queries are also listed to verify that the model-theoretic method here does generate the results we expect.

4 Central Query

The model-theoretic square of opposition sees the four corners as queries. One way to visualize the square is this:

263

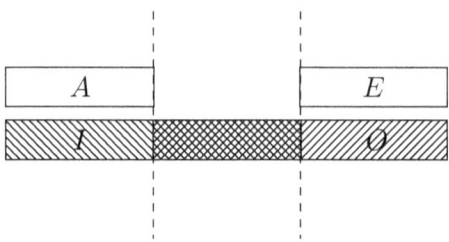

To say that A and E can both be false is to say that there are models that are included in neither query. There is, thus, a *central query* that makes up the gap between A and E; it also makes up the overlap between I and O.

Because (semantic) contraries are not unique, a single corner is not enough to generate a square. A universal query will, in general, have many disjoint queries, any one of which can create a square. This also allows the creation of more-complex polyhedra of opposition: any three disjoint queries can make a hexagon of opposition, any four an octagon, and so on.

But the model-theoretic approach also allows for a unique square given any corner. Because the antithesis of any universal query can be calculated, so too can the central query. Seen model-theoretically, there is a privileged square for any universal or existential query. This privileged square underlies the model-theoretic notion of duality. For TF, there are many contraries of $\phi \wedge \psi$, but only one with the complement $\phi \vee \psi$; for FOL_1, there are many contraries of $\forall x.\phi$, but only one with the complement $\exists x.\phi$.[11]

The central query of this privileged square is itself interesting. In FOL_1, the central query is not in general expressible in FOL_1. It will be defined by the conjunction of the I sentence and the O sentence— $\exists x.\phi \wedge \exists x.\neg\phi$, which is not in general equivalent to a sentence with a single quantifier. But the notion expressed here is a natural one; it is arguably a better match for the normal understanding of the English

---

[11]There is one situation in which there is a semantic square but not a model-theoretic square: Given $\bot$ or its complement $\top$. Because $\bot$ is never true—it is represented by a string of 0 that is 2^s long, for a language of length s—it has $2^{2^s} - 1$ contraries. Every sentence but $\top$ is a contrary of $\bot$, by the definition of semantic contraries.

word 'some', which seems to imply 'not all'. In the modal square, the central query is $\Diamond\phi \wedge \Diamond\neg\phi$, which is one possible reading of 'contingent'.[12]

5 Conclusion

Typically, the square of opposition is understood semantically, treating the relevant relations (contradictory, contrary, etc.) as dealing with the truth or falsity of the propositions. This understanding gives the squares a great deal of flexibility, for a given proposition with a contrary will typically have many. We now draw the square with contradictories diagonal, but Aristotle (at *Prior Analytics* 51b36–39) takes the contradictories to lie horizontally. With this organization, we can think of the various squares like pages of a book held with its spine at the top, bound together by a sentence and its contradictory but differing in the other corners.

But the model-theoretic approach gives a good approach to a privileged square, which is more stable, like a cornerstone. Each query that can have an antithesis has only one, and thus has only one square. This square is also syntactically privileged in logics that generate the square with the interplay of inner and outer negation. Flexibility is lost on this perspective, but there are compensating riches in the tightness of the connections.

References

[1] Jonathan Barnes, editor. *The Complete Works of Aristotle*. Princeton University Press, 1984.

[2] Ryan Christensen. The logic of $\triangle$. *Thought*, 2(4):1–7, 2014.

[3] Hilail Gildin. Aristotle and the moral square of opposition. *The Monist*, 54(1):100–105, 1970.

[4] Sebastian Löbner. Quantification as a major module of natural language semantics. In *Studies in discourse representation theory and the theory of generalized quantifiers*, pages 53–86. De Gruyter, 1986.

---

[12] It is also important in the logic of vagueness: 'It is indeterminate whether ϕ' is the central query of a square formed from the operator 'It is determinate that ϕ'. See [2].

[5] Stanley Peters and Dag Westerståhl. *Quantifiers in Language and Logic*. Clarendon Press, 2006.

[6] Hans Smessaert and Lorenz Demey. The unreasonable effectiveness of bitstrings in logical geometry. In Jean-Yves Béziau and Gianfranco Basti, editors, *The Square of Oppositions: A Cornerstone of Thought*, pages 197–214. Birkhäuser, 2017.

DISCUSSIVE LOGIC
A SHORT HISTORY OF THE FIRST
PARACONSISTENT LOGIC

FABIO DE MARTIN POLO
Ruhr-University Bochum
Department of Philosophy I
fabio.demartinpolo@rub.de

Abstract

In this paper we present an overview, with historical and critical remarks, of two articles by S. Jaśkowski ([20, 21] 1948 and [22, 23] 1949), which contain the oldest known formulation of a paraconsistent logic. Jaśkowski has built the logic – he termed *discussive* (D_2) – by defining two new connectives and by introducing a modal translation map from D_2 systems into Lewis' modal logic **S5**. Discussive systems, for their formal details and their original philosophical justification, have attracted discrete attention among experts. Indeed, in what follows, after having introduced Jaśkowski's methodology of building D_2 and his main philosophical motivations for providing such a system, we will explore some of the main contributions to the development of D_2.

1 The Origins of Discussive Logic

Throughout this paper we will consider the following classical connectives, $\sim$ *(negation),* $\wedge$ *(conjunction),* $\vee$ *(disjunction),* $\supset$ *(material implication), plus the modal operators,* $\Box$ *(necessary) and* $\Diamond$ *(possible). All*

I would like to thank Hitoshi Omori for reading and discussing with me several versions of this work. The preparation of this article was supported by a Sofja Kovalevskaja Award of the Alexander von Humboldt Foundation, funded by the German Ministry for Education and Research.

additions and changes will be explicitly stated and explained.
$\Gamma, \Delta, \Sigma, \ldots$ *and* $A, B, C, \ldots$ *denote sets of formulas and formulas, respectively.* $p, q, r \ldots$ *stand for propositional variables..*

1.1 The first discussive system

S. Jaśkowski (1906-1965)[1] is the author of several important logical and mathematical studies. To cite some of them, Jaśkowski is usually acknowledged as one of the inventors of the natural deduction calculus (accomplishing this work almost at the same time of G. Gentzen) and as the proponent of the first paraconsistent logic known as 'discussive' (or 'discursive') logic[2]. In [21] (which corresponds to the English translation of Jaśkowski's original article [20], published in 1948), the logician proposed a logic which should capture situations where discussants are in conflict. Jaśkowski's main idea was to consider a discussant's statement, p, as inherently consistent, but potentially incoherent with some other discussant's proposition. With this in mind, Jaśkowski focused his attention on a classically valid law, namely *ex contradictione quodlibet [sequitur]* ((ECQ), 'from a contradiction everything [follows]') – $p \supset (\sim p \supset q)$ – claiming that it should not be generally valid. His strategy, in order to invalidate (ECQ), has been that of getting rid of the classical connective of material implication, i.e., $\supset$, in favour of so-called 'discussive implication', i.e., $\to_d$. Lewis' modal logic **S5** has played a fundamental role in the formulation of such discussive systems, so, let's recall the definition of **S5**:

Definition 1.1. **S5** is axiomatized is follows:

If A is a theorem of **PC**, then A is a theorem of **S5**.

$\Box(A \supset B) \supset (\Box A \supset \Box B)$ \hfill (K)

$\Box A \supset A$ \hfill (T)

$\Diamond A \supset \Box \Diamond A$ \hfill (5)

and the following rules:

---

[1] For biographical informations one can consider [27, 16, 19]. For synthetic introductions to Jaśkowski's discussive logic, see, for example, [40, 42].

[2] Jaśkowski denoted this logic by $\mathbf{D}_2$, where the label '2' indicates that we are dealing with the 'two-valued discussive sentential calculus'.

$$\frac{A \quad A \supset B}{B} \text{ MP} \qquad \frac{A}{\Box A} \text{ Nec}$$

Finally, we say that a modal logic **L** is of **S5**-type iff $\mathbf{L} \subseteq \mathbf{S5}^3$.

Thanks to Lewis' modal system, Jaśkowski established the definition of discussive implication in the following way: $p \to_d q \stackrel{\text{def}}{=} \Diamond p \supset q$, validating thus the discussive version of *modus ponens*:

$$\frac{A \quad A \to_d B}{B} \text{ MP}_d$$

Additionally, we can get also the definition of 'discussive bi-implication', $p \leftrightarrow_d q \stackrel{\text{def}}{=} (\Diamond p \supset q) \wedge (\Diamond q \supset \Diamond p)$. Notice that, so defined, both, $\to_d$ and $\leftrightarrow_d$, are asymmetric connectives. One might wonder what the $\Diamond$ operator is meant to represent in a discussive framework. According to Jaśkowski's own perspective:

> To bring out the nature of the theses of such a system it would be proper to precede each thesis by the reservation: "in accordance with the opinion of one of the participants in the discussion" or "for a certain admissible meaning of the terms used". Hence the joining of a thesis to a discussive system has a different intuitive meaning than has assertion in an ordinary system. *Discussive assertion* includes an implicit reservation of the kind specified above, which [...] has its equivalent in $\Diamond$. [21, 43]

In a latest note, [23] (the English translation of the 1949 paper [22]), Jaśkowski proposed to substitute from the set of connectives also classical conjunction in favour of "discussive conjunction" and chose the following definition: $p \wedge_d q \stackrel{\text{def}}{=} p \wedge \Diamond q$. With this additional connective, then Jaśkowski defined again discussive bi-implication in the following manner: $p \leftrightarrow_d q \stackrel{\text{def}}{=} (p \to_d q) \wedge_d (q \to_d p)$. So, in sum, to prove discussive formulas, i.e., formulas including discussive connectives, Jaśkowski suggested to transform such formulas accordingly to their modal definitions and to prove the resulting modal formula in **S5**. In more rigorous terms:

---

[3] As known, **S5** has several equivalent axiomatization; for instance, one can employ (4) ($\Box A \supset \Box\Box A$) and (B) ($A \supset \Box\Diamond A$) instead of axiom (5).

Definition 1.2. $\mathbf{D_2}$ is the system whose language $\mathscr{L}$ includes the following set of connectives $\mathsf{S} = \{\sim, \vee, \wedge_d, \to_d, \leftrightarrow_d\}$. Take a function $\tau : \mathsf{Form}_{\mathbf{D_2}} \mapsto \mathsf{Form}_{\mathbf{S5}}$ such that, for any $A, B \in \mathsf{Form}_{\mathbf{D_2}}$:

$$\tau(p) = p$$
$$\tau(\sim A) = \sim \tau(A)$$
$$\tau(A \vee B) = \tau(A) \vee \tau(B)$$
$$\tau(A \wedge_d B) = \tau(A) \wedge \Diamond \tau(B)$$
$$\tau(A \to_d B) = \Diamond \tau(A) \supset \tau(B)$$
$$\tau(A \leftrightarrow_d B) = (\Diamond \tau(A) \supset \tau(B)) \wedge \Diamond(\Diamond \tau(B) \supset \tau(A))$$

Let $\Diamond \Gamma \stackrel{\text{def}}{=} \{\Diamond \tau(A_1), \ldots, \Diamond \tau(A_n) \mid A_1, \ldots, A_n \in \Gamma\}$, then for all $\Gamma \subseteq \mathsf{Form}_{\mathbf{D_2}}$ and $B \in \mathsf{Form}_{\mathbf{D_2}}$, we set:

$$\Gamma \models_{\mathbf{D_2}} B \text{ iff } \Diamond \Gamma \models_{\mathbf{S5}} \Diamond \tau(B).$$

In other words, a formula B is said to be a *discussive consequence* of a set of premises $\{A_1, \ldots, A_n\}$ just in case $\Diamond \tau(B)$ follows from the set $\{\Diamond \tau(A_1), \ldots, \Diamond \tau(A_n)\}$ in $\mathbf{S5}$. Following Jaśkowski:

> ... if a thesis A is recorded in a discussive system, its intuitive sense ought to be interpreted so as if it were preceded by the symbol $\Diamond$, that is, the sense: "it is possible that A". This is how an impartial arbiter might understand the theses of the various participants in the discussion. [21, 43]

The motivation behind this quote and Definition 1.2 can be intuitively explained with the following example. If we take formulas including $\to_d$ and replace it simply accordingly to τ we will obtain a great number of $\mathbf{S5}$ invalid formulas. In this case, even the identity, $A \to_d A$, if transformed in $\Diamond A \supset A$, turns out to be $\mathbf{S5}$-invalid. However, many of this negative results can be avoided, if we prefix $\Diamond$ to every modally translated formula. For example, $A \to_d A$, if translated as follows $\Diamond(\Diamond A \supset A)$, turns out to be $\mathbf{S5}$-valid.

Observation 1. To see the paraconsistent character of $\mathbf{D_2}$ consider that already in [21], the discussive version of (ECQ), $A \to_d (\sim A \to_d B)$, was not included as a theorem of $\mathbf{D_2}$. To see this, consider always the modal translation of (ECQ), i.e., $\Diamond(\Diamond A \supset (\Diamond {\sim} A \supset B))$, which is not valid in **S5**. Consequently to the rejection of (ECQ), the existence of contradictory statements, $\Diamond A$ and $\Diamond {\sim} A$, is possible without that their presence entails the 'overfilling' (triviality) of the system. However, the logic is not paraconsistent with respect to conjuncted contradictions, indeed, $\Diamond(\Diamond(A \wedge {\sim} A) \supset B)$ is still a theorem of **S5**. Moreover, notice that in this framework $\wedge$ adjunction fails (i.e., $A \wedge B$ cannot be inferred from A and B) and, for this specific reason, the $\{{\sim}, \vee, \wedge, \to_d\}$-fragment of $\mathbf{D}_2$ is usually classified among the *non-adjunctive* approaches to paraconsistent logics:

> ... discussive logic represents an ideology that is, to my mind, the most appropriate one for paraconsistency. To put it informally: at the very core of paraconsistency lies not negation, but conjunction. [...] With respect to inconsistency tolerating calculi, this connective seems to be the most important one. [45, 487]

Nonetheless, in [23], thanks to the presence of discussive conjunction, adjunction can be successfully restated in the system. The discussive version of the law of non contradiction (LNC), ${\sim}(A \wedge_d {\sim} A)$, remains a valid law. To see this consider always the **S5** invalid formula $\Diamond(A \wedge \Diamond {\sim} A)$. Finally, the discussive version of conjunctive (ECQ), $(A \wedge_d {\sim} A) \to_d B$, is no longer valid, making, thus, $\mathbf{D}_2$ paraconsistent also with respect to conjuncted contradictions.

Observation 2. Jaśkowski's definition of $\wedge_d$ and $\to_d$ are not the only ones available and, indeed, experts considered different variants, such as:

$$A \wedge_d^l B \stackrel{\text{def}}{=} \Diamond A \wedge B$$

$$A \wedge_d^s B \stackrel{\text{def}}{=} \Diamond A \wedge \Diamond B$$

$$A \to_d^s B \stackrel{\text{def}}{=} \Diamond A \supset \Diamond B$$

As one can easily see, the introduction of these new connectives tries to recover the asymmetry present in Jasśkowski's original proposal. Anyway, notice that the formulas $\Diamond(A \wedge \Diamond B)$, $\Diamond(\Diamond A \wedge B)$ and $\Diamond(\Diamond A \wedge \Diamond B)$

are all equivalent in **S5**, while $\Diamond(\Diamond A \supset B)$ and $\Diamond(\Diamond A \supset \Diamond B)$ are already equivalent in **S4** (a subset of **S5**). Moreover, as known since [21], $\mathbf{D}_2$ is a paraconsistent extension of the $\{\vee, \wedge, \supset\}$-fragment of classical logic. In other words, the discussive operators in $\mathbf{D}_2$ behave just like their classical counterparts. Interestingly, however, if we consider also an enriched language which includes a negation connective, the discussive logics generated by these new operators will no longer coincide with the $\{\sim, \vee, \wedge, \supset\}$-fragment of classical logic.

> It is not true thus that different translation clauses 'would have just the same consequences' [...]. Different choices of discussive conjunction and discussive implication would in fact define logics distinct from $\mathbf{D}_2$. [28, 215]

This is a struggling point. Indeed, as we will see in section 2.3, some notable problems arise in the formulation and comparison of axiomatic systems including different discussive connectives and negation.

1.2 Jaśkowski's Philosophical Motivations

In his celebrated *Metaphysics*, Aristotle claimed that 'the most indisputable of all beliefs is that contradictory statements are not at the same time true' ([3, Γ, 1011b13–14]), establishing, thus, – in a crystal clear way for the first time in the history of philosophy – one of the most celebrated and debated logical, psychological and ontological laws, i.e., the so-called law of non-contradiction (LNC). Roughly, Aristotle was convinced that the principle for which two opposite propositions, usually, one the negation of the other, cannot both be true at the same time had a very special status. Indeed, (LNC) corresponds, according to the Greek philosopher, to the most certain principle, which has a triple valence: it is a law of human rationality and reasoning (logic), it is a law governing reality (ontology) and, finally, it is a law concerning human beliefs (psychology). The discussions continued and, finally, during the middle ages, the debates on contradictions reached another fundamental turning point. An unknown author, usually acknowledged under the pseudonymous of Pseudo-Scotus, defined for the first time the principle of *ex contradictione quodlibet [sequitur]* in a commentary to Aristotle's *Analytica Priora* [43]. Importantly, William of Soissons, during the XII century, proposed the first known proof of the aforementioned principle and it is documented that already during the XIV century logicians knew about its existence and accepted (ECQ) as true[4]. However, the birth and the growing interest towards formal logical systems, strictly matched to philosophical considerations and objectives, has led some philosophers and logicians to re-consider also the validity and the truth of (LNC) and (ECQ). Jaśkowski has been among them. Indeed, in the first paragraphs of his celebrated 1949 article he develops a brief survey concerning the most important philosophical positions which, according to his reading, have provided some motivations to accept the presence of contradictory sentences (especially, Hegel and Marx)[5]. For instance, with respect to empirical sciences, Jaśkowski wrote:

> ... it is known that the evolution of the empirical disciplines is marked by periods in which the theorists are unable to explain the results of experiments by a homogenous and consistent theory, but use different hypotheses, which are not always consistent with one another, to explain the various groups of phenomena. This applies, for instance, to physics in its present-day stage. Some hypotheses are even termed "working" hypotheses when they result in certain correct predictions, but have no chance to be accepted for good, since they fail in some other cases. [21, 37]

The theoretical solution, according to Jaśkowski, is the following:

> ... we have to take into account the fact that in some cases we have to do with a system of hypotheses which, if subjected to a too consistent analysis, would result in a contradiction between themselves or with a certain accepted law, but which we use in a way that is restricted so as not to yield a self evident falsehood. [21, 37]

Indeed, in the paragraphs were he begins to elaborate more formally his ideas, Jaśkowski distinguishes very strictly between 'inconsistent' and

---

[4]Importantly, the works by William of Soissons have not been preserved, however a witness of his work is contained in John of Salisbury's *Metalogicon*.

[5]For more philosophical details on the consequences of adopting a paraconsistent point of view, one might consider [41].

'trivial' system. The first notion is linked to the presence, within the logical system under consideration, of two theses, one the negation of the other (p and $\sim p$); the second concept, instead, asserts that in a system it is possible to derive any formula if there is a couple of contradictory statements. So, as obvious, systems in which every proposition is derivable have no practical significance, since everything can be asserted. So, finally:

> ... the task is to find a system of the sentential calculus which: (1) when applied to the inconsistent systems would not always entail their overfilling, (2) would be rich enough to enable practical inference, (3) would have an intuitive justification. [21, 38]

Jaśkowski did not further elaborate his philosophical considerations, but, nowadays, scholars provided – by taking inspiration directly from Jaśkowski's brief suggestions – some interesting philosophical applications of $\mathbf{D}_2$ (for example, to the foundations of physical theories, to the notion of pragmatic (or partial) truth [10, 14], to the formal study of belief structures and argumentation schemes [17]).

2 The Development of Discussive Logic

Discussive systems have attracted discrete attention and various experts contributed to their development[6]. Our aim, in what follows, is to systematize and explain some of the main works concerning Jaśkowski's discussive logic. To keep the presentation as much as possible self-contained, we will restrict our attention to three distinct, even if connected, paths. More precisely, we will focus our attention on:

§2.1 the connections between discussive logic and modal systems;
§2.2 a family of logics, called "**J**" systems;
§2.3 the "direct" axiomatizations of $\mathbf{D}_2$, i.e., those systems which include axioms for discussive connectives.

---
[6] At the best of our knowledge, one previous attempt in that direction was made by Ciuciura in [4] from 1999. Nonetheless, in what follows, we wish to consider also alternative approaches towards discussive systems and enrich our considerations by commenting more recent works.

2.1 Connections to Modal Logics

2.1.1 Early developments

The tradition of modal studies connected to $\mathbf{D}_2$ started already in 1968 thanks to a paper by N. da Costa [11] and continued uninterrupted throughout the years. Roughly said:

> Besides non-adjunctiveness, another common obsession of discussivists concerns the alleged 'modal character' of $\mathbf{D}_2$. [28, 217]

Early remarkable results have been provided by J. Kotas in [24] from 1974. First of all, let's fix the next definition:

Definition 2.1. Let $\heartsuit \in \{\Box, \Diamond\}$. A $\heartsuit$-counterpart of a modal system $\mathbf{M}$ is defined as follows: $\heartsuit^n(\mathbf{M}) = \{A \mid \heartsuit^n A \in \mathbf{M}\}$, for $n \geq 1$.

With respect to Jaśkowski's $\mathbf{D}_2$, Kotas elaborated an axiomatization having as primitive connectives only $\sim$, $\supset$, $\Box$. We will denote this system $\mathbf{D}_2^K$, where 'K' stands for Kotas. The axioms of $\mathbf{D}_2^K$ are:

$$\Box(A \supset (\sim A \supset B)) \tag{K1}$$
$$\Box((A \supset B) \supset ((B \supset C) \supset (A \supset C))) \tag{K2}$$
$$\Box((\sim A \supset A) \supset A) \tag{K3}$$
$$\Box(\Box(A \supset B) \supset (\Box A \supset \Box B)) \tag{K4}$$
$$\Box(\Box A \supset A) \tag{K5}$$
$$\Box(\sim \Box A \supset \Box \sim \Box A) \tag{K6}$$
$$\text{Substitution} \tag{Sub}$$

$$\frac{\Box A \quad \Box(A \supset B)}{\Box B} \, \Box\text{MP} \qquad \frac{\Box A}{\Box\Box A} \, \text{R4} \qquad \frac{\Box A}{A} \, \text{Den} \qquad \frac{\sim\Box\sim A}{A} \, \text{Dep}\Box$$

As usual, if we want to add the possibility operator, we can define it: $\Diamond A \stackrel{\text{def}}{=} \sim\Box\sim A$. Notice that by having $\Diamond$ as a defined connective, Dep$\Box$ may be substituted by:

$$\frac{\Diamond A}{A} \, \text{Dep}$$

An important achievement of [24] is the presentation of the following equivalences between **S5**-type systems and various combinations of axioms and rules of $\mathbf{D}_2^K$:

K1-K6	(Sub)	($\Box$MP)	(R4)	(Den)	(Dep$\Box$)/(Dep)	Equivalent System
✓	✓	✓	✓	-	-	$\Box$**S5**
✓	✓	✓	✓	✓	-	**S5**
✓	✓	✓	✓	✓	✓	$\Diamond$**S5**

Notice that, according to the table above, Kotas proved that $\mathbf{D}_2^K$ is equivalent to $\Diamond$**S5**. This result allowed him, finally, to prove that $\mathbf{D}_2^K$ is finitely axiomatizable. To obtain his results, Kotas relied on two different Jaśkowski- style translation functions. Take τ of Definition 1.2 and substitute the clauses for $\wedge_d$ and $\rightarrow_d$ with the following ones:

$$\tau^*(A \wedge_d B) = \sim(\sim\tau^*(A) \vee \Box\sim\tau^*(B))$$
$$\tau^*(A \rightarrow_d B) = (\sim\Box\sim\tau^*(A) \supset \tau^*(B))$$

In addition, consider a map τ_1 such that $\mathsf{Form}_{\Diamond \mathbf{S5}} \mapsto \mathsf{Form}_{\mathbf{D}_2^K}$. For any $A, B \in \Diamond\mathbf{S5}$:

$$\tau_1(p) = p$$
$$\tau_1(\sim A) = \sim\tau_1(A)$$
$$\tau_1(A \supset B) = \sim\tau_1(A) \vee \tau_1(B)$$
$$\tau_1(\Box A) = \sim((\sim p \vee p) \wedge_d \tau_1(A))$$

First of all, the equivalence between $\mathbf{D}_2^K$ and $\Diamond$**S5** follows also thanks to the introduction of two additional connectives [24, 197], [46, 37], namely:

$$A \prec B \stackrel{\text{def}}{=} \Box(A \supset B) \qquad (\prec)$$
$$A \rightharpoonup B \stackrel{\text{def}}{=} \sim((\sim p \vee p) \wedge_d \sim(\sim A \vee B)) \qquad (\rightharpoonup)$$

In particular, Kotas showed that the interpretation τ turns the implication $\rightharpoonup$ in the strict implication $\prec$, and the interpretation τ_1 turns the implication $\prec$ in $\rightharpoonup$. Collecting all this together, Kotas proved that:

1. The translations maps τ and τ_1 establish that $\mathbf{D}_2^K$ and $\Diamond$**S5** are equivalent. In other words, if $\models_{\mathbf{D}_2^K} A$ then $\models_{\Diamond\mathbf{S5}} \tau(A)$ and if $\models_{\Diamond\mathbf{S5}} B$ then $\models_{\mathbf{D}_2^K} \tau_1(B)$, [24, 198-199].

2. $\mathbf{D}_2^K$ is a finitely axiomatizable system [24, 199].

Along these lines of studies, the polish logician T. Furmanowski [18] published a paper concerning the smallest modal system whose ◇- counterpart coincides with discursive logic. So, by starting from Kotas' axiomatization K1-K5, Furmanowski defined ◇**S4**, i.e., the ◇- counterpart of **S4**. As usual, by adding axiom K6 to the axiomatization, we get ◇**S5**. In particular, in [18], what's interesting, with respect to these systems, is the equality between ◇**S4** and ◇**S5**. This result is obtained by showing that both inclusions, (i) ◇**S4** ⊇ ◇**S5** and (ii) ◇**S5** ⊇ ◇**S4**, are satisfied. The latter inclusion is trivial since it is well-known that **S5** ⊇ **S4**. For (i), instead, we need to show that the axioms K1-K5 and the rules of inferences of [24] constitute a complete axiomatization of ◇**S4** ([18, 39]) and, secondly, to prove that the characteristic axiom of ◇**S5** K6 is also a formula of ◇**S4** ([18, 41]). This equality states that, for any A, $\models_{\diamond\mathbf{S4}} A$ just in case $\models_{\diamond\mathbf{S5}} A$. So, roughly, the quality of modality in ◇**S4** is the same as in ◇**S5**. From this result and the axiomatizations of ◇**S4** and ◇**S5**, Furmanowski proved that, for any system **S** such that, **S4** ⊆ **S** ⊆ **S5**: $\models_\mathbf{S} \diamond A$ if and only if $\models_{\diamond\mathbf{S5}} \diamond A$. At this point, with this background, Furmanowski defined Jaśkowski's discursive logic by starting from such a system **S**:

Definition 2.2. Let $\mathbf{D}(\mathbf{S})$ be a discursive system as based on a modal system **S**, such that **S4** ⊆ **S** ⊆ **S5**:

$$\mathbf{D}(\mathbf{S}) = \{A \in \mathsf{Form}_{\mathbf{D}(\mathbf{S})} \mid \diamond\tau(A) \in \mathbf{S}\}$$

Take Jaśkowski's translation map τ. Then: $\models_{\mathbf{D}(\mathbf{S})} A$ iff $\models_\mathbf{S} \diamond\tau(A)$.

Notice that, if $\mathbf{S} = \mathbf{S5}$, then $\mathbf{D}(\mathbf{S5}) = \mathbf{D}_2$. From this fact, and by the previous result for which, for any system **S4** ⊆ **S** ⊆ **S5**, it holds that $\models_\mathbf{S} \diamond A$ if and only if $\models_{\diamond\mathbf{S5}} \diamond A$, we may conclude that, for any such modal system **S**: $\mathbf{D}(\mathbf{S}) = \mathbf{D}_2$.

2.1.2 Recent developments

The tradition of modal studies connected to Jaśkowski's logic continued and largely increased. Recently, the gigantic work of M. Nasieniewski and A. Pietruszczak in [35, 36, 37] contributed to the development of the weakest regular modal logic[7] (denoted by $\mathbf{rS5}^M$) that defines $\mathbf{D}_2$. In [35], the authors analyse $\mathbf{S5}^M$, i.e., a normal modal logic presented previously by J. Perzanowski. Let $\mathbf{L}$ be any modal logic such that $\mathbf{L}$ defines $\mathbf{D}_2$ iff $\mathbf{D}_2 = \{A \in \mathsf{Form}_{\mathbf{D}_2} \mid \Diamond\tau(A) \in \mathbf{L}\}$. We denote with $\Diamond\mathbf{NS5}$ the set of all normal logics from $\Diamond\mathbf{S5}$. By having this in mind and by following the authors of [35], let's introduce the system $\mathbf{S5}^M$ with the following axioms:[8]

$$\Box p \supset \Diamond p \tag{D}$$
$$\Diamond\Box(\Diamond\Box p \supset \Box p) \tag{ML5}$$
$$\Diamond\Box(\Box p \supset p) \tag{MLT}$$

and the rule:

$$\frac{\Diamond\Diamond A}{\Diamond A} \; \mathrm{RM}_1^2$$

A preliminary result is that $\mathbf{S5}^M$ is the smallest logic in $\Diamond\mathbf{NS5}$ [35, 199] but, also, that $\mathbf{S5}^M$ is the smallest normal logic defining $\mathbf{D}_2$.

Starting from $\mathbf{S5}^M$, the authors consider $\mathbf{rS5}^M$, which is the smallest regular logic which contains (MLT) and (RM_1^2). As expected, $\mathbf{rS5}^M \in \Diamond\mathbf{RS5}$ and, moreover, it constitutes the smallest logic belonging to $\Diamond\mathbf{RS5}$. With respect to discussive logic, Nasieniewski and Pietruszczak aimed at showing that $\mathbf{rS5}^M$ is the smallest regular (non-normal) modal logic defining Jaśkowski's $\mathbf{D}_2$. To do this, the author of [36] consider again the function τ of Definition 1.2 together with the following map, labelled τ_2. Let τ_2 be a map such that $\mathsf{Form}_{\mathbf{rS5}^M} \mapsto \mathsf{Form}_{\mathbf{D}_2}$. For any formula $A, B \in \mathbf{rS5}^M$:

$$\tau_2(p) = p$$
$$\tau_2(\sim A) = \sim\tau_2(A)$$
$$\tau_2(A \vee B) = \tau_2(A) \vee \tau_2(B)$$

---

[7]As usual, we define a regular modal logic $\mathbf{L}$ as a set of modal formulas satisfying the following conditions: (i) $\mathbf{PC} \subseteq \mathbf{L}$, (ii) $\Diamond p \leftrightarrow \sim\Box\sim p \in \mathbf{L}$ and (iii) $\mathbf{L}$ is closed under *modus pones* for $\supset$, under the regularity rule $(A \wedge B) \supset C/(\Box A \wedge \Box B) \supset \Box C$, and under uniform substitution A/A', where A' is the result of uniform substitution of propositional variables in A. Moreover, $\mathbf{L}$ is said to be normal if $\mathrm{K} \in \mathbf{L}$ and $\mathrm{Nec} \in \mathbf{L}$.

[8]Notice that in all normal and regular modal logics axiom (D) can be equivalently formulated as $\Diamond(p \supset p)$.

$$\tau_2(A \wedge B) = \sim(\sim\tau_2(A) \vee \sim\tau_2(B))$$
$$\tau_2(A \supset B) = \sim\tau_2(A) \vee \sim\tau_2(B)$$
$$\tau_2(A \leftrightarrow B) = \sim(\sim(\sim\tau_2(A) \vee \tau_2(B)) \vee \sim(\sim\tau_2(B) \vee \tau_2(A)))$$
$$\tau_2(\Diamond A) = (p \vee \sim p) \wedge_d \tau_2(A)$$
$$\tau_2(\Box A) = \sim\tau_2(A) \rightarrow_d \sim(p \vee \sim p)$$

With this in mind, we are able to introduce $\mathbf{D_2}$ as follows:

Definition 2.3. Let $\mathbf{L}$ be any modal logic such that:

$$\mathbf{D(L)} \stackrel{\text{def}}{=} \{A \in \mathsf{Form}_{\mathbf{D_2}} \mid \Diamond\tau(A) \in \mathbf{L}\}$$

Then: $\mathbf{L}$ defines $\mathbf{D_2}$ iff $\mathbf{D(L)} = \mathbf{D_2}$.

So, for any modal logic $\mathbf{L}$ such that, if $\mathbf{L} \in \Diamond\mathbf{S5}$ then $\mathbf{L}$ defines $\mathbf{D_2}$. Additionally, $\mathbf{rS5^M} \in \Diamond\mathbf{RS5}$ and $\mathbf{S5^M} \in \Diamond\mathbf{NS5}$. For $\Diamond\mathbf{RS5}$ and $\Diamond\mathbf{NS5}$ being subsets of $\Diamond\mathbf{S5}$, we get that $\mathbf{rS5^M} \in \Diamond\mathbf{S5}$ and $\mathbf{S5^M} \in \Diamond\mathbf{S5}$. So, $\mathbf{rS5^M}$ and $\mathbf{S5^M}$ both define $\mathbf{D_2}$ and, hence, $\mathbf{D_2} = \mathbf{D(rS5^M)} = \mathbf{D(S5^M)}$. In other words, $\mathbf{rS5^M}$ is the regular version of the smallest normal modal logic $\mathbf{S5^M}$ such that (i) $\mathbf{rS5^M} \subsetneq \mathbf{S5^M}$ and (ii) every theorem beginning with $\Diamond$ of $\mathbf{rS5^M}$ is also a theorem of $\mathbf{S5^M}$ [35, 204]. So, finally, collecting together all these results, we get the main *desiderata* of [35]: $\mathbf{rS5^M}$ is the smallest regular non-normal modal logic defining $\mathbf{D_2}$.

Additionally, in [36], the authors showed that $\mathbf{rS5^M}$ can be axiomatized without the rule of inference (RM_1^2) and that it is the smallest regular logic which contains the following theorems:

$$\Box p \supset \Diamond\Box\Box p \tag{4_s}$$
$$\Box p \supset \Diamond\Box p \tag{5_c}$$

In other terms, $\mathbf{rS5^M} = \mathbf{C4_s5_c}$. Moreover, (i) if $\mathbf{rS5^M}$ contains (4_s) and (MLT), we get that $\mathbf{rS5^M} = \mathbf{C4_s(MLT)}$. Finally, (ii) $\mathbf{rS5^M} = \mathbf{C5_c(RM_1^2)}$ iff it contains (5_c) and is closed under the rule (RM_1^2) [36, 49].

In [37], Nasieniewski and Pietruszczak gave a Kripke semantics for the smallest regular modal logic $\mathbf{rS5^M}(= \mathbf{C4_s5_c})$. The paper contains specific frame conditions for $\mathbf{rS5^M}$ and completeness results. Let's begin with the next definition:

Definition 2.4. A frame for regular modal logic $\mathbf{rS5}^M (= \mathbf{C4_s5_c})$ is a triple $\mathcal{F}_{\mathbf{rS5}^M} = \langle W, \mathcal{R}, N \rangle$, where W is the set of *worlds*, $N \subseteq W$ consists of *regular worlds* and $\mathcal{R}$ is the accessibility relation[9]. Furthermore, $\mathcal{F}_{\mathbf{rS5}^M} = \langle W, \mathcal{R}, N \rangle$ satisfies the following conditions:

$$\forall w \in N, \exists u \in N(w\mathcal{R}u \wedge \forall x \in W(u\mathcal{R}x \Rightarrow w\mathcal{R}x)) \tag{Fr1}$$
$$\forall w \in N, \exists u \in N(w\mathcal{R}u \wedge \forall x \in W(\exists y \in N(u\mathcal{R}y \wedge y\mathcal{R}x) \Rightarrow w\mathcal{R}x)) \tag{Fr2}$$

($\mathbf{5_c}$) is valid in frames satisfying (Fr1) [37, 177] and ($\mathbf{4_s}$) is valid if the frame satisfies (Fr2) [37, 178]. Notice that both frame conditions constitute strengthening of seriality [37, 179]. Finally, as usual:

Definition 2.5. A model $\mathcal{M}_{\mathbf{rS5}^M} = \langle W, \mathcal{R}, N, v \rangle$ for $\mathbf{rS5}^M (= \mathbf{C4_s5_c})$ is based on a frame $\mathcal{F}_{\mathbf{rS5}^M}$ and on a valuation $v : \mathsf{Form}_{\mathbf{rS5}^M} \times W \to \{0,1\}$ such that for any $A \in \mathsf{Form}_{\mathbf{rS5}^M}$ and $w \in W$:

$$v(\Box A) = 1 \quad \text{iff} \quad w \in N \text{ and } \forall x \in \mathcal{R}(w),\ v(A,x) = 1$$
$$v(\Diamond A) = 0 \quad \text{iff} \quad w \notin N \text{ or } \exists x \in \mathcal{R}(w),\ v(A,x) = 1$$

where $\mathcal{R}(w) \stackrel{\text{def}}{=} \{x \in W \mid w\mathcal{R}x\}$.
A formula A is true in a model $\mathcal{M}_{\mathbf{rS5}^M}$ iff $v(A, w) = 1$ for any $w \in W$. A formula A is valid in a given frame $\mathcal{F}_{\mathbf{rS5}^M}$ iff it is true in all models $\mathcal{M}_{\mathbf{rS5}^M}$ based on the aforementioned frame.

In sum, the authors of [35, 36, 37] provided both an axiomatic system and a possible worlds semantics for the regular version of **S5** and, consequently, defined discussive logic on that formal basis[10]. From the perspective of Jaśkowski's $\mathbf{D_2}$, the work of Nasieniewski and Pietruszczak is interesting since it shows, not only that there other normal modal logics different from **S5** defining discussive logic, but that there are also non-normal regular versions of **S5** which define $\mathbf{D_2}$.

---

[9] If we let $W = N$, then we get the pair $\langle W, \mathcal{R} \rangle$, which corresponds to a frame for normal modal logics.

[10] Notice, finally, that we have restricted our attention just to some of the papers that Nasieniewski, Pietruszczak and collaborators devoted to $\mathbf{D_2}$. For more on their work see our conclusive remarks.

2.2 The 'J' Systems

Remarkably,

> ... [t]he year 1967 was a turning point in the development of the discussive logic. Newton C.A. da Costa and Lech Dubikajtis met in Paris and gradually commenced the development of the logic. [4, 10]

Indeed, as said above, in a paper from 1968 [11], da Costa and Dubikajtis presented the first modal-type axiomatization of D_2. The **S5**-type system they proposed, known as **J**, has become famous in the context of discussive systems. As remarked by the authors, **J** has several axiomatizations[11] and, in what follows, we will refer to the axiomatic system presented in [10] from 1995. Interestingly, **J**, and in particular some of its extensions, have been applied to philosophical problems, such as to the debate on the underlying logic of scientific theories. However, before turning to the philosophical applications of **J** and related systems, let's introduce them. **J** is the system composed by the following axioms and rules [10, 45]:

If A is a theorem of **S5**, then $\Box A$ is a theorem of **J**.

$$\frac{\Box A \quad \Box(A \supset B)}{\Box B} \Box\text{MP} \qquad \frac{\Box A}{A} \text{Den} \qquad \frac{\Diamond A}{A} \text{Dep} \qquad \frac{\Box A}{\Box\Box A} \text{R4}$$

J has been introduced in the literature as another $\Diamond$-counterpart of **S5** and, indeed, $\models_\mathbf{J} A$ iff $\models_\mathbf{S5} \Diamond A$. Starting from **J**, da Costa and Doria presented a first-order variant of it, denoted **J***, by adding the universal quantifier $\forall$ among the connectives. As usual, the existential quantifier can be defined $\exists x A \stackrel{\text{def}}{=} \sim\forall x \sim A$. Before, defining **J***, it is useful to recall the axiomatic system for **S5Q**$^=$ (quantified **S5** with identity):

> If A is a theorem of **PC**, then A is an theorem of **S5Q**$^=$.
>
> All axioms of **S5** (Definition 1.1), plus :
>
> $x = x$ \hfill (Id1)

---

[11] For other synthetic reconstructions one can also consider [4, 46].

$$x = y \supset (A(x) \leftrightarrow B(x)) \tag{Id2}$$
$$\forall x A(x) \supset A(t), \tag{$\forall$1}$$

where t is either a variable free for x in $A(x)$ or an individual constant. And the following rule:

$$\dfrac{A \supset B(x)}{A \supset \forall x B(x)} \; \text{R}\forall$$

Now, the language of **J*** coincides the language of **S5Q**$^=$ and, indeed, **J*** is introduced as follows:

> If A is an theorem of **S5Q**$^=$, then $\Box A$ is a theorem of **J***
>
> The axioms and rules of **J**, plus:

$$\dfrac{\Box(A \supset B(x))}{\Box(A \supset \forall x B(x))} \; \text{R}\Box\forall$$

where, in the rule (R$\Box\forall$), x is not free in A.

Notice that, differently from Jaśkowski's papers, da Costa and Doria considered left-discussive conjunction. Roughly, by adding both, $\wedge_d^l$ and $\to_d$, to **J** and **J***, the paraconsistent character of such systems. Indeed, the following formulas, governing the 'explosion' of logical systems, are not valid neither in **J** nor in **J***. Let $\wedge$ be classical conjunction:

$$A \to_d (B \to_d A \wedge B)$$
$$((A \wedge B) \to_d C) \to_d (A \to_d (B \to_d C))$$
$$A \to_d (\sim A \to_d B)$$
$$(A \to_d \sim A) \to_d B$$

Furthermore, let $\Gamma \stackrel{\text{def}}{=} \{A \mid \Gamma \vdash_{\textbf{J*}} A\}$. As usual, if Γ is the set of all formulas, then Γ is trivial. If not, Γ is non-trivial; if we have a formula A such that both $\Gamma \vdash_{\textbf{J*}} A$ and $\Gamma \vdash_{\textbf{J*}} \sim A$, then Γ is inconsistent. If not, Γ is consistent. With respect to these definitions, the two authors – who where interested in modelling situations in which scientists may reason through inconsistent sets of sentences, considered as "working hypothesis" [10, 46] – showed that their **J**-systems allow to deal with inconsistent and non-trivial sets of premises. In other words, da Costa and Doria proved that **J** and **J*** are paraconsistent logics.

2.2.1 D_2, J^* & the foundations of physics

Recall that Jaśkowski believed that "the evolution of the empirical disciplines is marked by periods in which [...] the results of experiments [...] are not always consistent with one another" [21]. Accordingly, the inconsistent results are to be considered as 'working hypothesis', i.e., as sentences that are taken *as if they were true* to inspect their respective consequences and establish which one describes more accurately scientific phenomena. da Costa and Doria tried to make sense of Jaśkowski's idea by elaborating a variant of J^* which can be used as underlying logic for physical theories. The starting point has been represented by the (formal) conceptions of physical structure and theory, due to M.L. Dalla Chiara and G. Toraldo di Francia (see [15, 10, 14]). First of all, a 'physical structure' $\mathcal{A}$ is a set-theoretic structure of the following form:

$$\mathcal{A} = \langle M, S, < Q_0, Q_1, \ldots, Q_n >, \rho \rangle$$

where, M represents a set of mathematical structures. Notice, the authors of [15] aimed at modelling physical concepts, such as vector spaces, as set-theoretic structures, by taking the axioms of **ZF** set theory. Secondly, S is a set of 'physical situations', i.e., a set of physical states assumed by a physical system in a certain time interval. In other words, S is the element of the physical structure that 'mirrors' the physical theory that $\mathcal{A}$ is trying to capture. Each Q_k ($0 \leq k \leq n$) is an 'operationally defined quantity' whose domain of definition is some $S_1 \subseteq S$. As a convention, let Q_0 denote time. To be clear, if we wish to measure a quantity Q_k of a physical system in a state $s \in S$ at a time t_k, with $1 \leq k \leq n$, we get an interval $I(k, t_k)$ of the real number line $\mathbb{R}$. So, if we measure time, i.e., Q_0, the result we obtain is a 'time interval'. t and t_k represent time instants and we express, in $\mathscr{L}$, the 'acceptable values' of Q_k at t_i as $q_k(t_i)$. So, in a certain sense, *all values* in a interval $I(k, t_k)$ are 'appropriate values' for the measurement of the quantity Q_k of the physical system in a state $s \in S$. Finally, ρ associates mathematical structures of M to their physically meaningful quantity.

To see how this framework is supposed to work, as usual, let $A(t, q_k(t_k))$ be a formula whose only free variables are those one expressing time instants, (t and t_k). Formally, $\models_s A(t, q_k(t_k))$ means that a formula A, in a certain interval of time, is true for a physical state s if there are values t^0 and q_k^0 (of Q_k) in the interval $I(t, t_k)$, with $1 \leq k \leq n$, such that $A(t^0, q_k^0)$ is true in s. Now, let $\models_{\mathcal{A}} A(t, q_k(t_k))$ denote that A is true in the physical structure $\mathcal{A}$. If we obtain t in I_t and q_k in $I(t, t_k)$, so that $\sim A(t, q_k(t_k))$ is also true in $\mathcal{A}$, then the physical theory captured by $\mathcal{A}$ is paraconsistent. In other words, as one might have expected, with respect to $\mathcal{A}$, we get a paraconsistent physical theory whenever $\models_{\mathcal{A}} A(t, q_k(t_k))$ and $\models_{\mathcal{A}} \sim A(t, q_k(t_k))$.[12] At this point, da Costa and Doria aimed at demonstrating that:

> ... the underlying logic of a physical theory in Dalla Chiara and di Francia approach is most adequately represented by Jaśkowski's discussive logic. [10, 57]

and, more precisely, by **J****. This system is similar to **J***, but imposes some more restrictions on bound variables [10, 14]. Take again **S5Q**$^=$ and let $\uplus A \stackrel{\text{def}}{=} \forall x_n A(x_n)$ be denoting a formula A preceded by a sequences of universal quantifiers so that all variables in A are bound. **J**** is constituted by the following axioms and rules:

If A is an instance of a theorem of **S5Q**$^=$, then $\Box \uplus A$ is a theorem of **J***.

$\Box \uplus (\Box(A \supset B) \supset (\Box A \supset \Box B))$ \hfill (J$_1^{**}$)

$\Box \uplus (\Box A \supset A)$ \hfill (J$_2^{**}$)

$\Box \uplus (\Diamond A \supset \Box \Diamond A)$ \hfill (J$_3^{**}$)

$\Box \uplus (\forall x A(x) \supset A(t))$ \hfill (J$_4^{**}$)

$\Box \uplus (x = x)$ \hfill (J$_5^{**}$)

---

[12] To be clear, consider the following example due to [14, 849-850]. Take Newton's second law: $F = m \cdot a$. The variables appearing in the equation corresponds to the physical quantities to be measured: 'force' (F), 'mass' (m) and 'acceleration' (a). If we take a state $s \in S$, their values stand in the following three intervals $I(F_1, F_2) \subseteq \mathbb{R}$, $I(m_1, m_2) \subseteq \mathbb{R}$ and $I(a_1, a_2) \subseteq \mathbb{R}$. When we are able to find three real numbers $p_1 \in I(F_1, F_2)$, $q_1 \in I(m_1, m_2)$ and $r_1 \in I(a_1, a_2)$, such that $p_1 = q_1 \cdot r_1$, then it holds that $\models_s F = m \cdot a$. Likewise, if we encounter the opposite situation, namely we find three real numbers, in their respective intervals, such that $p_2 \neq q_2 \cdot r_2$, also these three real numbers can be considered as acceptable values for solving the equation. So, $\models_s \sim(F = m \cdot a)$ and, hence, Newton's second law, in the very same physical situation s, is both, true and false. In this case, for the same situation s, Newton's law is a proposition C, such that $\models_s C$ and $\models_s \sim C$. However, $\models_s C \wedge \sim C$ does not hold, since it would mean to find three real numbers p, q, r, in their respective intervals, for which the conjunction $p = q \cdot r \wedge p \neq q \cdot r$ holds.

$$\Box \uplus (x = y \supset (A(x) \leftrightarrow A(y))) \tag{J_6^{**}}$$

$$\frac{\Box \uplus A \quad \Box \uplus (A \supset B)}{\Box \uplus B} \uplus\Box\text{MP} \quad \frac{\Box \uplus A}{A} \uplus\text{Den} \quad \frac{\Box \uplus A}{\Box \uplus \Box A} \uplus\text{R4}$$

$$\frac{\Diamond \uplus A}{A} \uplus\text{Dep} \quad \frac{\Box \uplus (A \supset B(x))}{\Box \uplus (A \supset \forall x B(x))} \text{R}\uplus\Box\forall$$

So: $\models_{\mathbf{J^{**}}} A$ iff $\models_{\mathbf{S5Q}} = \Diamond \uplus A$. Notice that vacuous quantification can be introduced/eliminated in any formula.

The only difference between $\mathbf{J^{**}}$ and $\mathbf{J^{*}}$ concerns the applications: the first one is more suitable than the second one to handle with, since there's no problem on the discussive interpretation of the free variables. Accordingly, a physical theory, denoted, $\mathcal{T}$, extends the notion of physical structure and, in sum, it is composed by the following elements:

1. A formal language $\mathscr{L}$.
2. A set of axioms $\mathscr{A}$ expressed in $\mathscr{L}$ such that $\mathscr{A} = \mathscr{A}_L \cup \mathscr{A}_M \cup \mathscr{A}_P$, where $\mathscr{A}_L, \mathscr{A}_M$ and $\mathscr{A}_P$ are, respectively, the set of logical, mathematical and physical axioms.
3. A language $\mathscr{L}_0 \subset \mathscr{L}$. The logic $\mathscr{L}_0$, used to deal with the mathematical structures of $\mathcal{T}$, is classical and, hence, $\mathscr{A}_M$ includes all classically valid formulas.
4. The axioms of $\mathbf{J^{**}}$ are included in $\mathscr{A}_L$ to deal with inconsistent sets of premises.
5. $\mathscr{A}_M$ must contain all axioms for the structures of M.
6. $\mathscr{A}_P$ contains all "physically motivated sentences".

So, finally, for A being a theorem of $\mathcal{T}$, then it holds that: if A is formulated in $\mathscr{L}_0$, then A is closed under classical consequence relation. Furthermore, from the perspective of inconsistent theories: for all $A \in \mathcal{T}$, A is closed under $\mathbf{J^{**}}$-consequence relation.

Notice that terms of $\mathscr{L}_0$ cannot refer to the quantities Q_k, but exclusively to mathematical structures of M, which are totally classical. More precisely, exactly the quantities Q_k induce the language to be paraconsistent. Indeed, if we are given a formula B such that its terms refer to some of the Q_k, generally, it can result that both, B and $\sim B$ are true in $\mathcal{T}$. Consequently, both sentences should be included in $\mathscr{A}_P$. Here's exactly the paraconsistent character of the definition of truth, i.e., in a physical theory $\mathcal{T}$, for some state $s \in S$ and a formula B, we can reach both, $\models_s B$ and $\models_s \sim B$. As said above, the acceptance of pairs of contradictory statements, such as B and $\sim B$, is meant to mirror those situations in which two inconsistent sentences are taken to be true with the aim to inspect their respective consequences and chose which one provides a more accurate description of the scientific phenomena under consideration. Of course, this does not mean that: $\models_s B \wedge \sim B$.

2.3 Introducing Discussive Connectives

In the previous discussion we have left apart the centrality of discussive connectives in formulating Jaśkowski's discussive logic in favour of an analysis principally focused on the development of the connections between $\mathbf{D}_2$ and modal systems. In what follows, we reverse the perspective by analysing some of the major attempts to give axioms to Jaśkowski's $\mathbf{D}_2$, without relying on translations and by considering directly a language including $\wedge_d$, $\to_d$ instead of $\wedge$, $\supset$. The challenge of providing such an axiomatization, usually known as 'Jaśkowski's problem' [46, 42], has been stated by N. da Costa already in 1975 [9, 14]:

> Is it possible to formulate a **natural** and **simple** axiomatization for $\mathbf{D}_2$ employing $\to_d, \wedge_d, \vee$ and $\sim$ as the only primitive connectives?

According to [25], the first non modal axiomatization of $\mathbf{D}_2$ has been proposed by Furmanowski but has never been published before Kotas' paper from 1975 [46]. It is worth having a look at Furmanowski's work not only for its historical importance, but also for the originality of the proposed axioms. Let $A, B, C, \ldots$ be formulas and let $\bot \stackrel{\text{def}}{=} \sim(A \vee \sim A)$. The discussive logic $\mathbf{D}_2^F$ is axiomatized by the following axioms:

$$\sim(A \supset (\sim A \supset B)) \to_d \bot \tag{F1}$$
$$(A \supset B) \supset ((B \supset C) \supset (A \supset C)) \to_d \bot \tag{F2}$$
$$\sim((\sim A \supset B) \supset A) \to_d \bot \tag{F3}$$
$$\sim((\sim A \supset B) \supset A) \to_d B \tag{F4}$$
$$\sim((\sim(A \supset B) \to_d \bot) \to_d ((\sim A \to_d \bot) \supset (\sim B \to_d \bot))) \to_d \bot \tag{F5}$$

$$\sim(\sim\sim(\sim A \supset \bot) \lor \sim\sim(\sim A \to_d \bot)) \to_d \bot \qquad \text{(F6)}$$
$$(\sim(A \supset B) \to_d C) \to_d ((\sim A \to_d C) \to_d (\sim B \to_d C)) \qquad \text{(F7)}$$
$$(\sim A \to_d \bot) \to_d A \qquad \text{(F8)}$$
$$(A \to_d B) \to_d (\sim(A \to_d B) \to_d B) \qquad \text{(F9)}$$
$$\sim(\sim\sim A \to_d B) \to_d A \qquad \text{(F10)}$$

Notice that, $\mathbf{D}_2^F$ is still 'impure' in the sense that, even though, Furmanowski did not include the modal operators, he still kept the presence of two conditionals, including the material one. So, strictly speaking, accordingly to [9], $\mathbf{D}_2^F$ cannot be regarded as a solution to 'Jaśkowski's problem'. In 1977 [12, 13] da Costa and Dubikajtis presented the first complete axiomatization of discussive logic including directly discussive connectives in the axiom schemata. In particular, da Costa and Dubikajtis [12] presented some axioms including $\to_d$ and $\wedge_d^l$. From now on, we will denote the discussive logic so formalized by $\mathbf{D}_2^l$, where 'l' indicates the presence of $\wedge_d^l$, instead of Jaśkowski's $\wedge_d$. So, the discussive logic $\mathbf{D}_2^l$ is axiomatized as follows

$$A \to_d (B \to_d A) \qquad \text{(Ax1)}$$
$$(A \to_d (B \to_d C)) \to_d ((A \to_d B) \to_d (A \to_d C)) \qquad \text{(Ax2)}$$
$$(A \wedge_d^l B) \to_d A \qquad \text{(Ax3)}$$
$$(A \wedge_d^l B) \to_d B \qquad \text{(Ax4)}$$
$$A \to_d (B \to_d (A \wedge_d^l B)) \qquad \text{(Ax5)}$$
$$A \to_d (A \lor B) \qquad \text{(Ax6)}$$
$$B \to_d (A \lor B) \qquad \text{(Ax7)}$$
$$(A \to_d C) \to_d ((B \to_d C) \to_d (A \lor B) \to_d C) \qquad \text{(Ax8)}$$
$$A \to_d \sim\sim A \qquad \text{(Ax9)}$$
$$\sim\sim A \to_d A \qquad \text{(Ax10)}$$
$$((A \to_d B) \to_d A) \to_d A \qquad \text{(Ax11)}$$
$$\sim(A \lor \sim A) \to_d B \qquad \text{(Ax12)}$$
$$\sim(A \lor B) \to_d \sim(B \lor A) \qquad \text{(Ax13)}$$
$$\sim(A \lor B) \to_d (\sim B \wedge_d^l \sim A) \qquad \text{(Ax14)}$$
$$\sim(\sim\sim A \lor B) \to_d \sim(A \lor B) \qquad \text{(Ax15)}$$

$$(\sim(A \vee B) \to_d C) \to_d ((\sim A \to_d B) \vee C) \quad \text{(Ax16)}$$
$$\sim((A \vee B) \vee C) \to_d \sim(A \vee (B \vee C)) \quad \text{(Ax17)}$$
$$\sim((A \to_d B) \vee C) \to_d (A \wedge_d^l \sim(B \vee C)) \quad \text{(Ax18)}$$
$$\sim((A \wedge_d^l B) \vee C) \to_d (A \to_d \sim(B \vee C)) \quad \text{(Ax19)}$$
$$\sim(\sim(A \vee B) \vee C) \to_d (\sim(\sim A \vee C) \vee \sim(\sim B \vee C)) \quad \text{(Ax20)}$$
$$\sim(\sim(A \to_d B) \vee C) \to_d (A \to_d \sim(\sim B \vee C)) \quad \text{(Ax21)}$$
$$\sim(\sim(A \wedge_d^l B) \vee C) \to_d (A \wedge_d^l \sim(\sim B \vee C)) \quad \text{(Ax22)}$$

$$\frac{A \quad A \to_d B}{B} \ \text{MP}_d$$

Remark 2.6. $\mathbf{D}_2^l$ includes the following set of connectives into its language $\{\sim, \vee, \wedge_d^l, \to_d\}$, where the only difference, as said, with $\mathbf{D}_2$ is the presence of left-discussive conjunction. Notice that, even though $\mathbf{D}_2^l$ constitutes a complete axiomatization [12, 54], from the perspective of [9], it might be still considered only as a 'partial' solution to 'Jaśkowski's problem'. Indeed, this time the 'impurity' of the axioms is not linked to the inclusion of other connectives than the discussive ones, plus $\sim$ and $\vee$, but to the presence of $\wedge_d^l$. As remarked above (Observation 2), the interaction of $\sim$ with different discussive operators defines logics distinct from Jaśkowki's $\mathbf{D}_2$. Indeed, strictly speaking, since Jaśkowski's $\mathbf{D}_2$ included right-discussive conjunction, $\mathbf{D}_2^l$ can be considered only as a variation of $\mathbf{D}_2$.

More recently, J. Alama and H. Omori [44] presented a complete and sound axiomatization for discussive logic, including Jaśkowski's right-discussive conjunction (denoted $\mathbf{D}_2^r$). The starting point of [44] are the axioms of $\mathbf{D}_2^l$. The only necessary change to get $\mathbf{D}_2^r$, is to drop Ax19 and Ax22 in favour of:

$$\sim((A \wedge_d B) \vee C) \to_d (B \to_d \sim(A \vee C)) \quad \text{(Ax19$'$)}$$
$$\sim(\sim(A \wedge_d B) \vee C) \to_d (\sim(\sim A \vee C) \wedge_d B) \quad \text{(Ax22$'$)}$$

Moreover, in the axioms involving conjunction, one simply needs to replace $\wedge_d^l$ with $\wedge_d$. Of course, Ax19 and Ax22 of $\mathbf{D}_2^l$ mirrored the behaviour of negated left-discussive conjunction. Ax19$'$ and Ax22$'$ absolve the same job, but with respect to right-discussive conjunction. Both axioms are $\mathbf{D}_2$-valid if and only if their modally translated versions are S5-valid, i.e., just in case the following formulas are valid in S5, accordingly to τ (see 1.2):

$$\Diamond(\Diamond\sim((A \land \Diamond B) \lor C) \supset (\Diamond B \supset \sim(A \lor C)))$$
$$\Diamond(\Diamond\sim(\sim(A \land \Diamond B) \lor C) \supset (\sim(\sim A \lor C) \land \Diamond B))$$

Following the changes proposed in [44], it seems that $\mathbf{D}_2^r$ accomplishes, at least, the task of finding a correct and complete axioms system for Jaśkowski's discussive logic. At this point, it might be naturally asked if $\mathbf{D}_2^r$ goes even further and gives a positive and definitive answer to 'Jaśkowski's problem'. Up to now it seems to be the best candidate.

We wish to strengthen this idea by considering briefly two other axiomatizations for $\mathbf{D}_2$, both elaborated by J. Ciuciura in [6, 8]. First of all, consider again a set of connectives including lef-discussive conjunction and the axiomatic system proposed in [6] (denoted $\mathbf{D}_2^C$). Take Ax1-Ax8, plus MP_d, of $\mathbf{D}_2^l$, and add the following axioms:

$A \lor (A \to_d B)$ (C1)

$A \to_d \sim(\sim(A \lor B) \land_d^l \sim B \land_d^l \sim A)$ (C2)

$\sim(\sim(A \lor B) \land_d \sim B \land_d^l \sim A) \to_d$
$\qquad \to_d \sim(\sim(A \lor B \lor C) \land_d^l \sim C \land_d^l \sim B \land_d^l \sim A)$ (C3)

$\sim(\sim(A \lor B \lor C) \land_d^l \sim C \land_d^l \sim B \land_d^l \sim A) \to_d$
$\qquad \to_d \sim(\sim(A \lor B \lor C) \land_d^l \sim B \land_d^l \sim C \land_d^l \sim A)$ (C4)

$\sim(\sim(A \lor B) \land_d^l \sim B \land_d^l \sim A) \to_d ((A \lor \sim B) \to_d A)$ (C5)

$\sim(\sim(A \lor B \lor C) \land_d^l \sim C \land_d^l \sim B \land_d^l \sim A) \to_d$
$\qquad \to_d ((A \lor B \lor \sim C) \to_d (A \lor B))$ (C6)

$\sim(\sim(A \lor B \lor C) \land_d^l \sim C \land_d^l \sim B \land_d^l \sim A) \to_d$
$\qquad \to_d (\sim(\sim(A \lor B \lor \sim C) \land_d^l \sim\sim C \land_d^l \sim B \land_d^l \sim A) \to_d \sim(\sim B \land_d^l \sim A))$ (C7)

$\sim(\sim A \land_d^l \sim B) \to_d (A \lor B)$ (C8)

$(A \lor \sim\sim B) \to_d (A \lor B)$ (C9)

$(A \lor B) \to_d (A \lor \sim\sim B)$ (C10)

289

As usual, the consequence relation $\vdash_{\mathbf{D}_2^C}$ is determined by the axioms Ax1-Ax8, C1-C10 and by the rule MP_d. Additionally, to prove soundness and completeness results, Ciuciura proposed a possible world semantics for $\mathbf{D}_2^C$, in which all elements are identical to those of Definition 2.5, except that $W = N$ and that we include the following clauses:

$$v(A \wedge_d^l B, w) = 1 \quad \text{iff} \quad \exists x \in \mathcal{R}(w), \ v(A, x) = 1 \text{ and } v(B, w) = 1$$
$$v(A \to_d B, w) = 1 \quad \text{iff} \quad \forall x \in \mathcal{R}(w), \ v(A, x) = 0 \text{ or } v(B, w) = 1$$

Since $\mathbf{D}_2^C$ relies on an equivalence relation between worlds, the accessibility relation may be not explicitly stated in the clauses. In any case, these changes will not affect their meaning, [6, 239-240.]. Importantly, Ciuciura aimed at proving soundness and completeness of $\mathbf{D}_2^C$, but, unfortunately, in [44, 1171], it was proved that in $\mathbf{D}_2^C$ there is (at least) one unprovable formula. The point is struggling, since the formula in question, i.e., $\sim(A \vee \sim A) \to_d B$, is valid according to the Jaśkowski-style translation τ of Definition 1.2. Consequently, one might naturally doubt whether $\mathbf{D}_2^C$ is, in some sense, an axiomatization of Jaśkowski's discussive logic in the sense of [9], given also the presence of $\wedge_d^l$ instead of $\wedge_d$. However, in an another paper [8], Ciuciura restated the presence of right-discussive conjunction among the connectives and provided an axiomatic system for it. We denote Ciuciura's second axiomatization by $\mathbf{D}_2^{C*}$. Take again Ax1-Ax8 (replacing $\wedge_d^l$ with $\wedge_d$ in Ax3, Ax4, Ax5) and MP_d of $\mathbf{D}_2^l$, plus the following axioms:

$$A \vee (A \to_d B) \tag{C1*}$$
$$\sim(\sim A \wedge_d \sim\sim A \wedge_d \sim(A \vee \sim A)) \tag{C2*}$$
$$\sim(\sim A \wedge_d \sim B \wedge_d \sim(A \vee B)) \to_d$$
$$\qquad \to_d \sim(\sim A \wedge_d \sim B \wedge_d \sim C \wedge_d \sim(A \vee B \vee C)) \tag{C3*}$$
$$\sim(\sim A \wedge_d \sim B \wedge_d \sim C \wedge_d \sim(A \vee B \vee C)) \to_d$$
$$\qquad \to_d \sim(\sim A \wedge_d \sim C \wedge_d \sim B \wedge_d \sim(A \vee C \vee B)) \tag{C4*}$$
$$\sim(\sim A \wedge_d \sim B \wedge_d \sim C \wedge_d \sim(A \vee B \vee C)) \to_d$$
$$\qquad \to_d ((A \vee B \vee \sim C) \to_d (A \vee B)) \tag{C5*}$$
$$\sim(\sim A \wedge_d \sim B) \to_d (A \vee B) \tag{C6*}$$
$$(A \vee (B \vee \sim B)) \to_d \sim(\sim A \wedge_d \sim(B \vee \sim B)) \tag{C7*}$$

As in the case of $\mathbf{D}_2^C$, to prove soundness and completeness, Ciuciura proposed a possible worlds semantics, but dropping out the clause for $\wedge_d^l$ in favour of the following one for $\wedge_d$:

$$v(A \wedge_d B, w) = 1 \quad \text{iff} \quad \exists x \in \mathcal{R}(w),\ v(A, w) = 1 \text{ and } v(B, x) = 1$$

Some criticism has been moved against Ciuciura's $\mathbf{D}_2^{C*}$. J. Alama [1] noticed that if we take the axioms Ax1-Ax22 of da Costa's and Dubikajtis' $\mathbf{D}_2^l$, in comparison to the ones of Ciuciura, we will get a troublesome situation: the two axiomatizations share some theses (Ax1-Ax8), while some others are respectively unprovable. Technically, if we encounter this situation, the two logics under considerations are said to be 'orthogonal'. In this specific case [1, 4-8]:

Proposition 1. $\mathbf{D}_2^{C*} \nvdash$ Ax9, Ax12, Ax13, Ax14, Ax15, Ax16, Ax17, Ax18, Ax19, Ax20, Ax21, Ax22.

At this point, consequently, it might be naturally asked whether $\mathbf{D}_2^{C*}$ corresponds to a restriction of $\mathbf{D}_2^l$. The answer is no, since there is (at least) one axiom of $\mathbf{D}_2^{C*}$ which is $\mathbf{D}_2^l$-unprovable [1, 11]:

Proposition 2. $\mathbf{D}_2^l \nvdash C5^*$

In sum, $\mathbf{D}_2^{C*}$. and $\mathbf{D}_2^l$, one with respect to the other, are not complete axiomatizations and, moreover, they ought to be called as orthogonal, i.e., they overlap and each one has theorems which are not formulas of the other. Finally, also the addition of new axioms still confirms that Ciuciura's axiomatization $\mathbf{D}_2^{C*}$ is an incomplete system of axioms [44, 1168.].

Notice, finally, that $\mathbf{D}_2^{C*}$ also fails to be an axiomatization Jaśkowski's $\mathbf{D}_2$, in the sense that there are $\mathbf{D}_2$-valid formulas, that are unprovable in $\mathbf{D}_2^{C*}$ [44, 1167-1170], namely[13]:

$$A \to_d \sim\sim A$$
$$\sim(A \vee \sim A) \to_d B$$
$$\sim(A \vee B) \to_d \sim(B \vee A)$$
$$\sim(\sim\sim A \vee B) \to_d \sim(A \vee B)$$

---

[13] Notice that those $\mathbf{D}_2^{C*}$ unprovable formulas correspond to Ax9, Ax12, Ax13 and Ax15 of both, $\mathbf{D}_2^r$ and $\mathbf{D}_2^l$.

Remark 2.7. In conclusion, all these considerations lead us in doubting that $\mathbf{D}_2^C$ and $\mathbf{D}_2^{C*}$ did provide a solution to 'Jaśkowksi's problem'. Furthermore, given the presence of both, Observation 2 and of Proposition 2, also $\mathbf{D}_2^l$ seems to be far from providing a solution. Nonetheless, $\mathbf{D}_2^r$, as elaborated in [44], seems to be an adequate candidate to settle positively the problem raised in [9].

3 Conclusive remarks

We have selected some of the perspectives under which discussive systems can be considered and, for the sake of brevity, we have chosen to explain and discuss just some of the main contributions present in the literature. For example, we have analysed how 'Jaśkowski's problem' might be solved, given the axiomatic systems we discussed. Nonetheless, many other works could have been considered (to cite a few of them, see [25, 9, 46, 30]). J. Perzanowski, in the critical notes to [23, 59], showed how to define 'discussive negation', i.e., $\sim_d A \stackrel{\text{def}}{=} \Diamond \sim A$. Interestingly, the equivalence between $\Diamond \sim A$ and $\sim \Box A$, makes, in fact, $\sim_d$ equal to 'un-necessity'. However, there are only few articles considering these kind of extensions of the set of discussive connectives. Remarkably, in [7], there's an axiomatization of discussive logic including also $\sim_d$ among the connectives, but, unfortunately, this attempt has some problems (see, [44, 1178-1179]). Hence, the challenge of developing an axiomatization for $\mathbf{D}_2$, including also $\sim_d$, is still open.

As remarked several times, Jaśkowski's logic has attracted discrete attention and many other research paths have been inaugurated. For instance, there has been some interest in developing discussive logic by getting rid of classical **S5**, in favour of other non-classical systems (see, among others [26, 5, 2]). Additionally, the work of connecting $\mathbf{D}_2$ to modal logics (especially, the articles by M. Nasieniewski and colleagues) increased (for example, [38, 31, 39]). Among their gigantic work, it's worth mentioning the proposal of an 'adaptive' (inconsistency-tolerant) version of discussive logic (see [32, 33, 34] and [29]).

From a more philosophical perspective, instead, one might find another interesting application of discussive logic to the philosophy of sciences in [10], where, in addition to the applications of $\mathbf{J}^{**}$ to the foundations of physical theories, the authors propose also a theory of 'pragmatic' (or 'partial') truth. The intuition underlying their idea, roughly, is that, with respect to inconsistent informations, scientists work with such informations *as if they were true*, and do not take them to be true *simpliciter*. Also in this case, **J*** and **J**** show their usefulness in modelling reasoning with inconsistent sets of premises. Importantly, in [17], the authors – by taking inspiration from Jaśkowski's main motivation to build $\mathbf{D}_2$ – propose a four-valued discussive logic ($\mathbf{D}_4$) with the aim of capturing situations in which discussants put forward inconsistent opinions. Roughly, this work includes a 'doxastic' variant of discussive logic, allows to distinguish among different agents, each one with its respective set of beliefs, and models (through a function) the agents' capabilities (e.g., perception, expert-supplied knowledge, communication, discussion). The idea is that a reasoner, that starts from a lack of informations, can – in the process of acquiring more data – reach either support or refutation of such data. However, if there's an overload of informations, the reasoner may reach both, truth and falsity, i.e., inconsistent data.

In conclusion, as said, this overview is not exhaustive and, indeed, our aim was to indicate just some of the most interesting directions that discussive logic oriented researches have taken, by starting from Jaśkowski's papers. We think that thanks to its historical importance as the first known formulation of a paraconsistent logic and to its subsequent developments, discussive logic is still an interesting and vital field of investigation.

References

[1] J. Alama, "Some problems with two axiomatizations of discussive logic", 2014 (arXiv:1403.7777).

[2] S. Akama, J. Abe, and K. Nakamatsu, "Constructive discursive logic with strong negation," *Logique et Analyse*, vol. 54, no. 215, pp. 395–408, 2011.

[3] Aristotle, "Metaphysics", in *The Basic Works of Aristotle*, (R. McKeon, ed., New Introduction by C.D.C. Reeve) The Modern Library: NY, 2001 (Originally published by Random House: NY).

[4] J. Ciuciura, "History and development of the discursive logic," *Logica Trianguli*, vol. 3, pp. 3–31, 1999.

[5] J. Ciuciura, "Intuitionistic discursive system (*IDS*)," *Bulletin of the Section of Logic*, vol. 29, no. 1/2, pp. 57–62, 2000.

[6] J. Ciuciura, "On the da Costa, Dubikajtis and Kotas' system of discursive logic $\mathbf{D}_2^*$," *Logic and Logical Philosophy*, no. 14, pp. 235–252, 2005.

[7] J. Ciuciura, "A quasi-discursive system $\mathbf{ND}_2^+$," *Notre Dame Journal of Formal Logic*, vol. 47, no. 3, pp. 371–384, 2006.

[8] J. Ciuciura, "Frontiers of the discursive logic," *Bulletin of the Section of Logic*, vol. 37, no. 2, pp. 81–92, 2008.

[9] N. da Costa, "Remarks on Jaśkowski's discussive logic," *Reports on Mathematical Logic*, vol. 4, pp. 7–16, 1975.

[10] N. da Costa and F. Doria, "On Jaśkowski's discussive logics," *Studia Logica*, no. 54, pp. 33–60, 1995.

[11] N. da Costa and L. Dubikajtis, "Sur la logique discoursive de Jaśkowski," *Bulletin Acad. Polonaise des Sciences Math., Astr. et Phys*, vol. 16, pp. 551–557, 1968.

[12] N. da Costa and L. Dubikajtis, "On Jaśkowski's discussive logics," in *Non-classical Logics, Model Theory, and Computability* (A. Arruda, ed.), pp. 37–56, North-Holland Publishing Company: NY, 1977.

[13] N. da Costa and J. Kotas, "On some modal logical systems defined in connexion with Jaśkowski's problem," in *Non-classical Logics, Model Theory, and Computability* (A. Arruda, ed.), pp. 57–72, North-Holland Publishing Company: NY, 1977.

[14] N. da Costa, D. Krause, and O. Bueno, "Paraconsistent logics and paraconsistency," in *Philosophy of logic* (D. Jacquette, ed.), pp. 791–911, Elsevier, 2007.

[15] M. Dalla Chiara and G. Toraldo di Francia, *Le teorie fisiche. Un'analisi formale.* Bollati Boringhieri: Torino, 1981.

[16] L. Dubikajtis, "The life and works of Stanisław Jaśkowski," *Studia Logica*, vol. 34, no. 2, pp. 109–116, 1975.

[17] B. Dunin-Kęplicz, A. Powała, and A. Szałas, "Variations on Jaśkowski's discursive logic," in *The Lvov-Warsaw School. Past and Present* (Á. Garrido and U. Wybraniec-Skardowska, eds.), pp. 485–497, Birkhäuser: Cham, 2018.

[18] T. Furmanowski, "Remarks on discussive propositional calculus," *Bulletin of the Section of Logic*, vol. 4, no. 1, pp. 33–36, 1975.

[19] A. Indrzejczak, "Stanisław Jaśkowski: Life and work," in *The Lvov-Warsaw School. Past and Present* (Á. Garrido and U. Wybraniec-Skardowska, eds.), pp. 457–464, Birkhäuser: Cham, 2018.

[20] S. Jaśkowski, "Rachunek zdan dla systemów dedukcyjnych sprzecznych",

Studia Societatis Scientiarum Torunensis, Sectio A, no. 5, pp. 57–77, 1948.

[21] S. Jaśkowski, "A propositional calculus for inconsistent deductive systems," *Logic and Logical Philosophy*, vol. 7, pp. 35–56, 1999. English Translation of [20] by O. Wojtasiewicz and J. Perzanowski.

[22] S. Jaśkowski, "O koniunkcji dyskusyjnej w rachunku zdań dla systemów dedukcyjnych sprzecznych," *Studia Societatis Scientiarum Torunensis, Sectio A*, no. 8, pp. 171–172, 1949.

[23] S. Jaśkowski, "On the discussive conjunction in the propositional calculus for inconsistent deductive systems," *Logic and Logical Philosophy*, vol. 7, pp. 57–59, 1999. English Translation of [22] by O. Wojtasiewicz and J. Perzanowski.

[24] J. Kotas, "The axiomatisation of Jaśkowski's discussive system," *Studia Logica*, vol. 33, no. 2, pp. 195–200, 1974.

[25] J. Kotas, "Discussive sentential calculus of Jaśkowski," *Studia Logica*, vol. 34, no. 2, pp. 149–168, 1975.

[26] J. Kotas and N. da Costa, "A new formulation of discussive logic," *Studia Logica*, vol. 38, no. 4, pp. 429–225, 1979.

[27] J. Kotas and A. Pieczkowski, "Scientific works of Stanisław Jaśkowski," *Studia Logica*, vol. 21, pp. 7–15, 1967.

[28] J. Marcos, "Modality and paraconsistency", *The Logica Yearbook 2004* (M. Bilkova and L. Behounek, eds.), pp. 213–222, Filosofia: Prague, 2005.

[29] J. Meheus, "An adaptive logic based on Jaśkowskis approach to paraconsistency," *Journal of Philosophical Logic*, vol. 35, no. 6, pp. 539–567, 2006.

[30] K. Mruczek-Nasieniewska and M. Nasieniewski, "A Kotas-style characterisation of minimal discussive logic," *Axioms*, vol. 8, no. 4 108, pp. 1–17, 2019.

[31] K. Mruczek-Nasieniewska, M. Nasieniewski, and A. Pietruszczak, "A modal extension of Jaśkowski's discussive logic," *Logic Journal of the IGPL*, vol. 27, no. 4, pp. 451–477, 2019.

[32] M. Nasieniewski, "A comparison of two approaches to parainconsistency: Flemish and polish," *Logic and Logical Philosophy*, vol. 9, no. 9, pp. 47–74, 2001.

[33] M. Nasieniewski, "The axiom of Mckinsey-Sobociński $K1$ in the framework of discussive logics," *Logique et Analyse*, pp. 315–324, 2003.

[34] M. Nasieniewski, "An adaptive logic based on Jaśkowski's logic D_2," *Logique et analyse*, vol. 47, no. 185/188, pp. 287–304, 2004.

[35] M. Nasieniewski and A. Pietruszczak, "The weakest regular modal logic defining Jaskowski's logic D_2," *Bulletin of the Section of Logic*, vol. 37, no. 3/4, pp. 197–210, 2008.

[36] M. Nasieniewski and A. Pietruszczak, "New axiomatizations of the weakest regular modal logic defining Jaskowski's logic $\mathbf{D}_2$," *Bulletin of the Section of Logic*, vol. 38, no. 1/2, pp. 45–50, 2009.

[37] M. Nasieniewski and A. Pietruszczak, "Semantics for regular logics connected with Jaskowski's discussive logic $\mathbf{D}_2$," *Bulletin of the Section of Logic*, vol. 38, no. 3/4, pp. 173–187, 2009.

[38] A. Nasieniewski, M.and Pietruszczak, "A method of generating modal logics defining jaśkowski's discussive logic $\mathbf{D}_2$," *Studia Logica*, vol. 97, no. 1, pp. 161–182, 2011.

[39] M. Nasieniewski and A. Pietruszczak, "Axiomatisations of minimal modal logics defining Jaśkowski-like discussive logics," in *Trends in Logic XIII. Gentzen's and Jaśkowski's Heritage. 80 Years of Natural Deduction and Sequent Calculi* (A. Indrzejczak, J. Kaczmarek, and M. Zawidzki, eds.), pp. 149–163, Łódź University Press: Łódź, 2014.

[40] G. Priest, "Introduction: Paraconsistent logics," *Studia Logica*, vol. 43, no. 1/2, pp. 3–16, 1984.

[41] G. Priest, J. Beall, and B. Armour-Garb, *The Law of Non-contradiction. New Philosophical Essays*. Clarendon Press: Oxford (UK), 2006.

[42] G. Priest, K. Tanaka, and Z. Weber, "Paraconsistent logic," in *Stanford Encyclopedia of Philosophy* (E. Zalta, ed.), 2018.

[43] Pseudo-Scotus, "Questions on Aristotle's Prior Analytics. Question X", in *Medieval Formal Logic. Obligations, Insolubles and Consequences* (M. Yrjönsuuri, ed.), Kluwer: Dordrecht, 2001, pp. 225-234.

[44] H. Omori and J. Alama, "Axiomatizing Jaśkowski's discussive logic $\mathbf{D}_2$," *Studia Logica*, vol. 106, no. 6, pp. 1163–1180, 2018.

[45] M. Urchs, "On the role of adjunction in para(in)consistent logic", *Paraconsistency: The logical way to the inconsistent* (Carnielli, W., Coniglio, A. M. and D'ottaviano, I. M., eds.), pp. 487–499, CRC Press, 2002.

[46] V. Vasyukov, "A new axiomatization of Jaśkowski's discussive logic," *Logic and logical Philosophy*, vol. 9, no. 9, pp. 35–46, 2001.

www.ingramcontent.com/pod-product-compliance
Lightning Source LLC
Chambersburg PA
CBHW050129170426
43197CB00011B/1770